Advances in MATHEMATICAL ECONOMICS

Aims and Scope. The project is to publish *Advances in Mathematical Economics* once a year under the auspices of the Research Center for Mathematical Economics. It is designed to bring together those mathematicians who are seriously interested in obtaining new challenging stimuli from economic theories and those economists who are seeking effective mathematical tools for their research.

The scope of *Advances in Mathematical Economics* includes, but is not limited to, the following fields:

- Economic theories in various fields based on rigorous mathematical reasoning.
- Mathematical methods (e.g., analysis, algebra, geometry, probability) motivated by economic theories.
- Mathematical results of potential relevance to economic theory.
- Historical study of mathematical economics.

Authors are asked to develop their original results as fully as possible and also to give a clear-cut expository overview of the problem under discussion. Consequently, we will also invite articles which might be considered too long for publication in journals.

Shigeo Kusuoka • Toru Maruyama
Editors

Advances in Mathematical Economics

Volume 18

Springer

Editors
Shigeo Kusuoka
Professor
Graduate School of Mathematical Sciences
The University of Tokyo
3-8-1 Komaba, Meguro-ku
Tokyo 153-8914, Japan

Toru Maruyama
Professor
Department of Economics
Keio University
2-15-45 Mita, Minato-ku
Tokyo 108-8345, Japan

ISSN 1866-2226 1866-2234 (electronic)
ISBN 978-4-431-56241-2 978-4-431-54834-8 (eBook)
DOI 10.1007/978-4-431-54834-8
Springer Tokyo Heidelberg New York Dordrecht London

Printed on acid-free paper

Springer is part of Springer Science+Business Media (www.springer.com)

Table of Contents

Research Articles

Charles Castaing, Christiane Godet-Thobie, Le Xuan Truong, and Bianca Satco
Optimal Control Problems Governed by a Second Order Ordinary Differential Equation with *m*-Point Boundary Condition 1

Shigeo Kusuoka and Yusuke Morimoto
Stochastic Mesh Methods for Hörmander Type Diffusion Processes 61

Survey Article

Alexander J. Zaslavski
Turnpike Properties for Nonconcave Problems 101

Note

Yuhki Hosoya
A Characterization of Quasi-concave Function in View of the Integrability Theory 135

Subject Index 141

Instructions for Authors 143

Adv. Math. Econ. 18, 1–59 (2014)

Advances in
MATHEMATICAL ECONOMICS

Optimal Control Problems Governed by a Second Order Ordinary Differential Equation with *m*-Point Boundary Condition

Charles Castaing[1], **Christiane Godet-Thobie**[2], **Le Xuan Truong**[3], **and Bianca Satco**[4]

[1] Département de Mathématiques de Brest, Case 051, Université Montpellier II, Place E. Bataillon, 34095 Montpellier cedex, France (e-mail: charles.castaing@gmail.com)

[2] Laboratoire de Mathématiques de Brest, CNRS-UMR 6205, Université de Bretagne Occidentale, 6, avenue Le Gorgeu, CS 93837, 29238 Brest Cedex 3, France (e-mail: Christiane.godet-thobie@univ-brest.fr)

[3] Department of Mathematics and Statistics, University of Economics of HoChiMinh City, 59C Nguyen Dinh Chieu Str. Dist. 3, HoChiMinh City, Vietnam (e-mail: lxuantruong@gmail.com)

[4] Stefan cel Mare University of Suceava, Suceava, Romania (e-mail: bisatco@eed.usv.ro)

Received: August 22, 2013
Revised: November 20, 2013

JEL classification: C61, C73

Mathematics Subject Classification (2010): 34A60, 34B15, 47H10, 45N05

Abstract. Using a new Green type function we present a study of optimal control problem where the dynamic is governed by a second order ordinary differential equation (SODE) with *m*-point boundary condition.

Key words: Differential game, Green function, *m*-Point boundary, Optimal control, Pettis, Strategy, Sweeping process, Viscosity

1. Introduction

The pioneering works concerning control systems governed by second order ordinary differential equations (SODE) with three point boundary condition are developed in [2, 16]. In this paper we present some new applications of the Green function introduced in [11] to the study of viscosity problem in Optimal Control Theory where the dynamic is governed by (SODE) with m-point boundary condition. The paper is organized as follows. In Sect. 2 we recall and summarize the properties of a new Green function (Lemma 2.1) with application to a second order differential equation with m-point boundary condition in a separable Banach space E of the form

$$(SODE)\begin{cases} \ddot{u}_{\tau,x,f}(t) + \gamma \dot{u}_{\tau,x,f}(t) = f(t),\ t \in [\tau, 1] \\ u_{\tau,x,f}(\tau) = x,\ u_{\tau,x,f}(1) = \sum_{i=1}^{m-2} \alpha_i u_{\tau,x,f}(\eta_i). \end{cases}$$

Here γ is positive, $f \in L^1_E([0, 1])$, m is an integer number > 3, $0 \leq \tau < \eta_1 < \eta_2 < \cdots < \eta_{m-2} < 1$, $\alpha_i \in \mathbf{R}$ $(i = 1, 2, \ldots, m-2)$ satisfying the condition

$$\sum_{i=1}^{m-2} \alpha_i - 1 + \exp(-\gamma(1-\tau)) - \sum_{i=1}^{m-2} \alpha_i \exp(-\gamma(\eta_i - \tau)) \neq 0 \tag{1.1.1}$$

and $u_{\tau,x,f}$ is the trajectory $W^{2,1}_E([\tau, 1])$-solution to (SODE) associated with $f \in L^1_E([0, 1])$ starting at the point $x \in E$ at time $\tau \in [0, 1[$. By Lemma 2.1, $u_{\tau,x,f}$ and $\dot{u}_{\tau,x,f}$ are represented, respectively, by

$$\begin{cases} u_{\tau,x,f}(t) = e_{\tau,x}(t) + \int_0^1 G_\tau(t,s) f(s)ds,\ \forall t \in [\tau, 1] \\ \dot{u}_{\tau,x,f}(t) = \dot{e}_{\tau,x}(t) + \int_0^1 \frac{\partial G_\tau}{\partial t}(t,s) f(s)ds,\ \forall t \in [\tau, 1] \end{cases}$$

where G_τ is the Green function defined in Lemma 2.1 with

$$\begin{cases} e_{\tau,x}(t) = x + A_\tau(1 - \sum_{i=1}^{m-2} \alpha_i)(1 - \exp(-\gamma(t-\tau)))x,\ \forall t \in [\tau, 1] \\ \dot{e}_{\tau,x}(t) = \gamma A_\tau \left(1 - \sum_{i=1}^{m-2} \alpha_i\right) \exp(-\gamma(t-\tau))x,\ \forall t \in [\tau, 1] \\ A_\tau = \left(\sum_{i=1}^{m-2} \alpha_i - 1 + \exp(-\gamma(1-\tau)) - \sum_{i=1}^{m-2} \alpha_i \exp(-\gamma(\eta_i - \tau))\right)^{-1}. \end{cases}$$

We stress that both existence and uniqueness and the integral representation formulas of solution and its derivative for (SODE) via the new Green function are of importance of this work. Indeed this allows to treat several new applications to optimal control problems and also some viscosity solutions for the value function governed by (SODE) with m-point boundary condition. In Sect. 3, we treat an optimal control problem governed by (SODE) in a separable Banach space

$$(SODE)_\Gamma \begin{cases} \ddot{u}_f(t) + \gamma \dot{u}_f(t) = f(t),\ f \in S^1_\Gamma \\ u_f(0) = x, \quad u_f(1) = \sum_{i=1}^{m-2} \alpha_i u_f(\eta_i) \end{cases}$$

where Γ is a measurable and integrably bounded convex compact valued mapping and S^1_Γ is the set of all integrable selections of Γ. We show the compactness of the solution set and the existence of optimal control for the problem

$$\begin{cases} \ddot{u}_f(t) + \gamma \dot{u}_f(t) = f(t),\ f \in S^1_\Gamma \\ u_f(0) = x, \quad u_f(1) = \sum_{i=1}^{m-2} \alpha_i u_f(\eta_i), \end{cases}$$

$$\inf_{f \in S^1_\Gamma} \int_0^1 J(t, u_f(t), \dot{u}_f(t), \ddot{u}_f(t))dt.$$

These results lead naturally to the problem of viscosity for the value function associated with this class of (SODE) which is presented in Sect. 4. In Sect. 5 we deal with a class of (SODE) with Pettis integrable second member. Existence and compactness of the solution set are also provided. Open problems concerning differential game governed by (SODE) and (ODE) with strategies are given in Sect. 6. We finish the paper by providing an application to the dynamic programming principle (DPP) and viscosity property for the value function associated with a sweeping process related to a model in Mathematical Economics [25].

2. Existence and Uniqueness

Let E be a separable Banach space. We denote by E^* the topological dual of E; $\overline{B}_E$ is the closed unit ball of E; $\mathcal{L}([0, 1])$ is the σ algebra of Lebesgue measurable sets on $[0, 1]$; $\lambda = dt$ is the Lebesgue measure on $[0, 1]$; $\mathcal{B}(E)$ is the σ algebra of Borel subsets of E. By $L^1_E([0, 1])$, we denote the space of all Lebesgue–Bochner integrable E-valued functions defined on $[0, 1]$. Let

$C_E([0,1])$ be the Banach space of all continuous functions $u : [0,1] \to E$ endowed with the sup-norm and let $C^1_E([0,1])$ be the Banach space of all functions $u \in C_E([0,1])$ with continuous derivative, endowed with the norm

$$\max\left\{\max_{t\in[0,1]} \|u(t)\|\,,\ \max_{t\in[0,1]} \|\dot{u}(t)\|\right\}.$$

We also denote $W^{2,1}_E([0,1])$ the space of all continuous functions in $C_E([0,1])$ such that their first derivatives are continuous and their second weak derivatives belong to $L^1_E([0,1])$.

We recall and summarize a new Green type function given in [11] that is a key ingredient in the statement of the problems under consideration.

Lemma 2.1. *Let $0 \le \tau < \eta_1 < \eta_2 < \cdots < \eta_{m-2} < 1$, $\gamma > 0$, $m > 3$ be an integer number, and $\alpha_i \in \mathbf{R}$ $(i = 1, \ldots, m-2)$ satisfying the condition*

$$\sum_{i=1}^{m-2} \alpha_i - 1 + \exp(-\gamma(1-\tau)) - \sum_{i=1}^{m-2} \alpha_i \exp(-\gamma(\eta_i - \tau)) \neq 0. \qquad (1.1.1)$$

Let E be a separable Banach space and let $G_\tau : [\tau,1] \times [\tau,1] \to \mathbf{R}$ be the function defined by

$$G_\tau(t,s) = \begin{cases} \frac{1}{\gamma}(1-\exp(-\gamma(t-s)))\,, & \tau \le s \le t \le 1 \\ 0, & \tau \le t < s \le 1 \end{cases}$$
$$+\frac{A_\tau}{\gamma}(1-\exp(-\gamma(t-\tau)))\,\phi_\tau(s), \qquad (2.1)$$

where

$$\phi_\tau(s) = \begin{cases} 1-\exp(-\gamma(1-s)) - \sum_{i=1}^{m-2} \alpha_i\,(1-\exp(-\gamma(\eta_i - s)))\,, & \tau \le s < \eta_1 \\ 1-\exp(-\gamma(1-s)) - \sum_{i=2}^{m-2} \alpha_i\,(1-\exp(-\gamma(\eta_i - s)))\,, & \eta_1 \le s \le \eta_2 \\ \ldots\ldots & \\ 1-\exp(-\gamma(1-s)), & \eta_{m-2} \le s \le 1, \end{cases} \qquad (2.2)$$

and

$$A_\tau = \left(\sum_{i=1}^{m-2} \alpha_i - 1 + \exp(-\gamma(1-\tau)) - \sum_{i=1}^{m-2} \alpha_i \exp(-\gamma(\eta_i - \tau))\right)^{-1}. \tag{2.3}$$

Then the following assertions hold

(i) For every fixed $s \in [\tau, 1]$, *the function* $G_\tau(.,s)$ *is right derivable on* $[\tau, 1[$ *and left derivable on* $]\tau, 1]$. *Its derivative is given by*

$$\left(\frac{\partial G_\tau}{\partial t}\right)_+ (t,s) = \begin{cases} \exp(-\gamma(t-s)), & \tau \le s \le t < 1 \\ 0, & \tau \le t < s < 1 \end{cases}$$
$$+A_\tau \exp(-\gamma(t-\tau))\phi_\tau(s), \tag{2.4}$$

$$\left(\frac{\partial G_\tau}{\partial t}\right)_- (t,s) = \begin{cases} \exp(-\gamma(t-s)), & \tau \le s < t \le 1 \\ 0, & \tau < t \le s \le 1 \end{cases} + A_\tau \exp(-\gamma(t-\tau))\phi_\tau(s). \tag{2.5}$$

(ii) $G_\tau(\cdot,\cdot)$ *and* $\frac{\partial G_\tau}{\partial t}(\cdot,\cdot)$ *satisfies*

$$|G_\tau(t,s)| \le M_{G_\tau} \text{ and } \left|\frac{\partial G_\tau}{\partial t}(t,s)\right| \le M_{G_\tau}, \quad \forall (t,s) \in [\tau,1] \times [\tau,1],$$

where

$$M_{G_\tau} = \max\{\gamma^{-1}, 1\}\left[1 + |A_\tau|\left(1 + \sum_{i=1}^{m-2} |\alpha_i|\right)\right].$$

(iii) If $u \in W_E^{2,1}([\tau, 1])$ *with* $u(\tau) = x$ *and* $u(1) = \sum_{i=1}^{m-2} \alpha_i u(\eta_i)$, *then*

$$u(t) = e_{\tau,x}(t) + \int_\tau^1 G_\tau(t,s)(\ddot{u}(s) + \gamma \dot{u}(s))ds, \quad \forall t \in [\tau, 1],$$

where

$$e_{\tau,x}(t) = x + A_\tau(1 - \sum_{i=1}^{m-2} \alpha_i)(1 - \exp(-\gamma(t-\tau)))x.$$

(iv) Let $f \in L_E^1([\tau, 1])$ *and let* $u_f : [\tau, 1] \to E$ *be the function defined by*

$$u_f(t) = e_{\tau,x}(t) + \int_\tau^1 G_\tau(t,s) f(s)ds, \quad \forall t \in [\tau, 1].$$

Then we have

$$u_f(\tau) = x, \quad u_f(1) = \sum_{i=1}^{m-2} \alpha_i u_f(\eta_i).$$

Further the function u_f *is derivable on* $[\tau, 1]$ *and its derivative* $\dot{u}_f$ *is defined by*

$$\dot{u}_f(t) = \lim_{h\to 0} \frac{u_f(t+h) - u_f(t)}{h} = \dot{e}_{\tau,x}(t) + \int_\tau^1 \frac{\partial G_\tau}{\partial t}(t,s) f(s)ds,$$

with

$$\dot{e}_{\tau,x}(t) = \gamma A_\tau (1 - \sum_{i=1}^{m-2} \alpha_i) \exp(-\gamma(t-\tau))x.$$

(v) *If* $f \in L^1_E([\tau, 1])$, *the function* $\dot{u}_f$ *is scalarly derivable, and its weak derivative* $\ddot{u}_f$ *satisfies*

$$\ddot{u}_f(t) + \gamma \dot{u}_f(t) = f(t) \quad a.e. \quad t \in [\tau, 1].$$

Proof. (i) Let $s \in [\tau, 1]$ and $t \in [\tau, 1]$. We consider two following cases.

Case 1 $t \neq s$. For every small $h > 0$ with $h < \min\{|t-s|, 1-t\}$, we have

$$\frac{G_\tau(t+h,s) - G_\tau(t,s)}{h} = \begin{cases} (\gamma h)^{-1} \exp(-\gamma(t-s))(1 - \exp(-\gamma h)), \\ \quad \tau \le s < t < 1 \\ 0, \\ \quad \tau \le t < s \le 1 \end{cases}$$
$$+ A_\tau \exp(-\gamma(t-\tau))$$
$$\times (1 - \exp(-\gamma h))(\gamma h)^{-1} \phi_\tau(s).$$

Hence $G_\tau(\cdot, s)$ is right derivable at $t \in [\tau, 1[\setminus \{s\}$ and

$$\left(\frac{\partial G_\tau}{\partial t}\right)_+ (t,s) = \begin{cases} \exp(-\gamma(t-s)), & \tau \le s < t < 1 \\ 0, & \tau \le t < s \le 1 \end{cases}$$
$$+ A_\tau \exp(-\gamma(t-\tau)) \phi_\tau(s).$$

Similarly, it is not difficult to check that $G_\tau(\cdot, s)$ is left derivable at $t \in]\tau, 1] \setminus \{s\}$ and

$$\left(\frac{\partial G_\tau}{\partial t}\right)_- (t,s) = \begin{cases} \exp(-\gamma(t-s)), & \tau \le s < t \le 1 \\ 0, & \tau < t < s \le 1 \end{cases}$$
$$+ A_\tau \exp(-\gamma(t-\tau)) \phi_\tau(s).$$

Case 2 $t = s$. Given $0 < h < 1 - s$. We have

$$\frac{G_\tau(t+h,s) - G_\tau(t,s)}{h} = (\gamma h)^{-1}(1 - \exp(-\gamma h)) + A_\tau \exp(-\gamma(t-\tau))(1 - \exp(-\gamma h)) \times (\gamma h)^{-1}\phi_\tau(s),$$

hence

$$\left(\frac{\partial G_\tau}{\partial t}\right)_+ (s,s) = 1 + A_\tau \exp(-\gamma(s-\tau))\phi_\tau(s).$$

Now given $0 < h < s - \tau$. We have

$$\frac{G_\tau(t-h,s) - G_\tau(t,s)}{h} = A_\tau \exp(-\gamma(t-\tau)) \times (1 - \exp(-\gamma h))(\gamma h)^{-1}\phi_\tau(s),$$

hence

$$\left(\frac{\partial G_\tau}{\partial t}\right)_+ (s,s) = A_\tau \exp(-\gamma(s-\tau))\phi_\tau(s).$$

(ii) It is easy to see that $|\phi_\tau(s)| \leq 1 + \sum_{i=1}^{m-2} |\alpha_i|$ for all $s \in [0,1]$. So, from the definition of G_τ we deduce that for all $s, t \in [\tau, 1]$

$$|G_\tau(t,s)| \leq \frac{1}{\gamma}\left[1 + |A_\tau|\left(1 + \sum_{i=1}^{m-2} |\alpha_i|\right)\right] \leq M_{G_\tau}.$$

Similarly we deduce that for all $s, t \in [\tau, 1]$

$$\left|\frac{\partial G_\tau}{\partial t}(t,s)\right| \leq 1 + |A_\tau|\,|\phi_\tau(s)| \leq 1 + |A_\tau|\left(1 + \sum_{i=1}^{m-2} |\alpha_i|\right) \leq M_{G_\tau}.$$

(iii) Let $x^* \in E^*$. By definition of G_τ, for all $t \in [\tau, 1]$, we have

$$\left\langle x^*, \int_\tau^1 G_\tau(t,s)\ddot{u}(s)ds \right\rangle = \int_\tau^1 \langle x^*, G_\tau(t,s)\ddot{u}(s)\rangle ds$$
$$= \frac{1}{\gamma}\int_\tau^t (1 - \exp(-\gamma(t-s)))\langle x^*, \ddot{u}(s)\rangle ds$$
$$+ \frac{A_\tau}{\gamma}(1 - \exp(-\gamma(t-\tau)))\int_\tau^1 \langle x^*, \phi_\tau(s)\ddot{u}(s)\rangle ds.$$

On the other hand

$$\int_\tau^t (1-\exp(-\gamma(t-s)))\,\langle x^*, \ddot{u}(s)\rangle ds$$
$$= (\exp(-\gamma(t-\tau))-1)\,\langle x^*, \dot{u}(\tau)\rangle + \gamma \int_\tau^t \exp(-\gamma(t-s))\langle x^*, \dot{u}(s)\rangle ds$$

and $\int_\tau^1 \langle x^*, \phi_\tau(s)\ddot{u}(s)\rangle ds = I_1 + I_2$ where

$$I_1 = \sum_{i=1}^{m-1} \int_{\eta_i-1}^{\eta_i} (1-\exp(-\gamma(1-s)))\,\langle x^*, \ddot{u}(s)\rangle ds$$
$$= (\exp(-\gamma(1-\tau))-1)\,\langle x^*, \dot{u}(\tau)\rangle + \gamma \int_\tau^1 \exp(-\gamma(1-s))\langle x^*, \dot{u}(s)\rangle ds$$
$$I_2 = -\sum_{i=1}^{m-2}\sum_{j=i}^{m-2} \alpha_j \int_{\eta_i-1}^{\eta_i} (1-\exp(-\gamma(\eta_j-s)))\langle x^*, \dot{u}(s)\rangle ds$$
$$= -\sum_{i=1}^{m-2} \alpha_i\,(\exp(-\gamma(\eta_i-\tau))-1)\,\langle x^*, \dot{u}(\tau)\rangle$$
$$-\gamma \sum_{i=1}^{m-2}\sum_{j=i}^{m-2} \int_{\eta_{i-1}}^{\eta_i} \exp(-\gamma(\eta_i-s))\langle x^*, \dot{u}(s)\rangle ds$$

with $\eta_0 = \tau$, $\eta_{m-1} = 1$.
Hence

$$\left\langle x^*, \int_\tau^1 G_\tau(t,s)(\ddot{u}(s)+\gamma\dot{u}(s))ds \right\rangle$$
$$= \frac{1}{\gamma}(\exp(-\gamma(t-\tau))-1)\langle x^*, \dot{u}(\tau)\rangle$$
$$+ \frac{A_\tau}{\gamma}(1-\exp(-\gamma(t-\tau)))\langle x^*, \dot{u}(\tau)\rangle$$
$$\times \left[\exp(-\gamma(1-\tau)) - 1 - \sum_{i=1}^{m-2} \alpha_i\,(\exp(-\gamma(\eta_i-\tau))-1)\right]$$
$$+ \int_\tau^t \langle x^*, \dot{u}(s)\rangle ds + A_\tau\,(1-\exp(-\gamma t)) \sum_{i=1}^{m-2}\left(1-\sum_{j=i}^{m-2}\alpha_j\right)$$
$$\times \int_{\eta_{i-1}}^{\eta_i} \langle x^*, \dot{u}(s)\rangle ds.$$

This implies that

$$\langle x^*, \int_0^1 G_\tau(t,s)(\ddot{u}(s)+\gamma\dot{u}(s))ds\rangle = \langle x^*, u(t)-e_{\tau,x}(t)\rangle, \quad \forall t \in [\tau, 1].$$

Since this equality holds for every $x^* \in E^*$, we get

$$u(t) = e_{\tau,x}(t) + \int_\tau^1 G_\tau(t,s)(\ddot{u}(s)+\gamma\dot{u}(s))ds, \quad \forall t \in [\tau, 1].$$

(iv) Let $f \in L^1_E([0,1])$ and $u_f(t) = e_{\tau,x}(t) + \int_\tau^1 G_\tau(t,s)f(s)ds$, $\forall t \in [0,1]$. Then, by definition of G_τ in (i), we have $u_f(\tau) = x$ and

$$\begin{aligned}
u_f(1) &= e_{\tau,x}(1) + \frac{1}{\gamma}\int_\tau^1 (1-\exp(-\gamma(1-s)))f(s)ds \\
&\quad + \frac{A_\tau}{\gamma}(1-\exp(-\gamma(1-\tau)))\int_\tau^1 \phi_\tau(s)f(s)ds \\
&= e_{\tau,x}(1) + \frac{1}{\gamma}\int_\tau^1 \left[1-\exp(-\gamma(1-s))-\phi_\tau(s)\right]f(s)ds \\
&\quad + \frac{1}{\gamma}\left[A_\tau(1-\exp(-\gamma(1-\tau)))+1\right]\int_\tau^1 \phi_\tau(s)f(s)ds \\
&= e_{\tau,x}(1) + \frac{1}{\gamma}\sum_{i=1}^{m-2}\sum_{j=i}^{m-2}\alpha_j\int_{\eta_i-1}^{\eta_i}(1-\exp(-\gamma(\eta_i-s)))f(s)ds \\
&\quad + \frac{A_\tau}{\gamma}\sum_{i=1}^{m-2}\alpha_i(1-\exp(-\gamma(\eta_i-\tau)))\int_\tau^1 \phi_\tau(s)f(s)ds \\
&= e_{\tau,x}(1) + \frac{1}{\gamma}\sum_{i=1}^{m-2}\alpha_i\int_0^{\eta_i}(1-\exp(-\gamma(\eta_i-s)))f(s)ds \\
&\quad + \frac{A_\tau}{\gamma}\sum_{i=1}^{m-2}\alpha_i(1-\exp(-\gamma(\eta_i-\tau)))\int_\tau^1 \phi_\tau(s)f(s)ds.
\end{aligned}$$

From the definition of $e_{\tau,x}(t)$ and A_τ, we deduce that

$$\begin{aligned}
e_{\tau,x}(1) &= x + A_\tau\left(1-\sum_{i=1}^{m-2}\alpha_i\right)(1-\exp(-\gamma(1-\tau)))\,x \\
&= A_\tau\left[A_\tau^{-1}+1-\exp(-\gamma(1-\tau))+\sum_{i=1}^{m-2}\alpha_i(\exp(-\gamma(1-\tau))-1)\right]x \\
&= A_\tau\left[\sum_{i=1}^{m-2}\alpha_i\exp(-\gamma(1-\tau))-\sum_{i=1}^{m-2}\alpha_i\exp(-\gamma(\eta_i-\tau))\right]x
\end{aligned}$$

and

$$e_{\tau,x}(\eta_i) = x + A_\tau \left(1 - \sum_{j=1}^{m-2} \alpha_j\right) (1 - \exp(-\gamma(\eta_i - \tau)))\, x$$

$$= A_\tau \left[A_\tau^{-1} + 1 - \exp(-\gamma(\eta_i - \tau)) - \sum_{j=1}^{m-2} \alpha_j + \exp(-\gamma(\eta_i - \tau)) \sum_{j=1}^{m-2} \alpha_j \right] x$$

$$= A_\tau \Bigg[\exp(-\gamma(1 - \tau)) - \exp(-\gamma(\eta_i - \tau)) + \exp(-\gamma(\eta_i - \tau)) \sum_{j=1}^{m-2} \alpha_j$$

$$+ \sum_{j=1}^{m-2} \alpha_j \exp(-\gamma(\eta_j - \tau)) \Bigg] x.$$

Hence we deduce that

$$\sum_{i=1}^{m-2} \alpha_i e_{\tau,x}(\eta_i)$$

$$= A_\tau \Bigg[\sum_{i=1}^{m-2} \alpha_i \exp(-\gamma(1 - \tau)) - \sum_{i=1}^{m-2} \alpha_i \exp(-\gamma(\eta_i - \tau))$$

$$+ \left(\sum_{j=1}^{m-2} \alpha_j\right) \sum_{i=1}^{m-2} \alpha_i \exp(-\gamma(\eta_i - \tau)) - \left(\sum_{i=1}^{m-2} \alpha_i\right)$$

$$\times \sum_{j=1}^{m-2} \alpha_j \exp(-\gamma(\eta_j - \tau)) \Bigg] x = e_{\tau,x}(1).$$

So, by combining the above relations, we get

$$u_f(1) = \sum_{i=1}^{m-2} \alpha_i e_{\tau,x}(\eta_i) + \frac{1}{\gamma} \sum_{i=1}^{m-2} \alpha_i \int_0^{\eta_i} (1 - \exp(-\gamma(\eta_i - s))) f(s) ds$$

$$+ \frac{A_\tau}{\gamma} \sum_{i=1}^{m-2} \alpha_i (1 - \exp(-\gamma(\eta_i - \tau))) \int_\tau^1 \phi_\tau(s) f(s) ds$$

$$= \sum_{i=1}^{m-2} \alpha_i \Bigg[e_{\tau,x}(\eta_i) + \frac{1}{\gamma} \int_0^{\eta_i} (1 - \exp(-\gamma(\eta_i - s))) f(s) ds$$

$$+\frac{A_\tau}{\gamma}(1-\exp(-\gamma(\eta_i-\tau)))\int_\tau^1 \phi_\tau(s)f(s)ds\Bigg]$$

$$=\sum_{i=1}^{m-2}\alpha_i u_f(\eta_i).$$

On the other hand, by the same arguments as in [2] we can conclude that u_f is derivable and its derivative $\dot{u}_f$ is defined by

$$\dot{u}_f(t)=\dot{e}_{\tau,x}(t)+\int_\tau^1 \frac{\partial G}{\partial t}(t,s)f(s)ds,\quad \forall t\in[0,1].$$

(v) Indeed, let $t\in[0,1]$. Using the expression of $\frac{\partial G}{\partial t}$ in (i) we have

$$\dot{u}_f(t)=\dot{e}_{\tau,x}(t)+\int_\tau^t \exp(-\gamma(t-s))f(s)ds$$
$$+A_\tau\exp(-\gamma(t-\tau))\int_\tau^1 \phi_\tau(s)f(s)ds.$$

Whence

$$\langle x^*,\ddot{u}_f(t)\rangle=\frac{d}{dt}\langle x^*,\dot{u}_f(t)\rangle$$
$$=\langle x^*,\ddot{e}_{\tau,x}(t)\rangle+\frac{d}{dt}\int_\tau^t \exp(-\gamma(t-s))\langle x^*,f(s)\rangle ds$$
$$-A_\tau\gamma\exp(-\gamma(t-\tau))\int_\tau^1 \langle x^*,\phi_\tau(s)f(s)\rangle ds$$
$$=\langle x^*,\ddot{e}_{\tau,x}(t)\rangle+\langle x^*,f(t)\rangle-\gamma\int_\tau^t \exp(-\gamma(t-s))\langle x^*,f(s)\rangle ds$$
$$-A_\tau\gamma\exp(-\gamma(t-\tau))\int_\tau^1 \langle x^*,\phi_\tau(s)f(s)\rangle ds.$$

We also note that $\ddot{e}_{\tau,x}(t)=-\gamma\dot{e}_{\tau,x}(t)$. Therefore

$$\langle x^*,\ddot{u}_f(t)\rangle=\langle x^*,f(t)\rangle-\langle x^*,\gamma\dot{u}_f(t)\rangle.$$

This implies that $\dot{u}_f$ is scalarly derivable and

$$\ddot{u}_f(t)+\gamma\dot{u}_f(t)=f(t)\quad a.e.\quad t\in[0,1].$$

□

The following result is a direct application of Lemma 2.1.

Lemma 2.2. *With the notations of Lemma 2.1, assume* $0 \le \tau < \eta_1 < \eta_2 < \cdots < \eta_{m-2} < 1$, $\gamma > 0$, $m > 3$ *be an integer number, and* $\alpha_i \in \mathbf{R}$ $(i = 1, \ldots, m-2)$ *and (1.1.1). Let* $f \in C_E([\tau, 1])$ *(resp.* $f \in L^1_E([\tau, 1])$*). Then the m-point boundary problem*

$$\begin{cases} \ddot{u}_{\tau,x,f}(t) + \gamma \dot{u}_{\tau,x,f}(t) = f(t), \ t \in [\tau, 1] \\ u_{\tau,x,f}(\tau) = x, \ u_{\tau,x,f}(1) = \sum_{i=1}^{m-2} \alpha_i u_{\tau,x,f}(\eta_i) \end{cases}$$

has a unique $C^2_E([\tau, 1])$*-solution (resp.* $W^{2,1}_E([\tau, 1])$*-solution) which is given by the integral representation formulas*

$$\begin{cases} u_{\tau,x,f}(t) = e_{\tau,x}(t) + \int_\tau^1 G_\tau(t,s) f(s) ds, \ t \in [\tau, 1] \\ \dot{u}_{\tau,x,f}(t) = \dot{e}_{\tau,x}(t) + \int_\tau^1 \frac{\partial G_\tau}{\partial t}(t,s) f(s) ds, \ t \in [\tau, 1] \end{cases}$$

where

$$\begin{cases} e_{\tau,x}(t) = x + A_\tau (1 - \sum_{i=1}^{m-2} \alpha_i)(1 - \exp(-\gamma(t-\tau)))x, \\ \dot{e}_{\tau,x}(t) = \gamma A_\tau \left(1 - \sum_{i=1}^{m-2} \alpha_i\right) \exp(-\gamma(t-\tau))x, \\ A_\tau = \left(\sum_{i=1}^{m-2} \alpha_i - 1 + \exp(-\gamma(1-\tau)) - \sum_{i=1}^{m-2} \alpha_i \exp(-\gamma(\eta_i - \tau))\right)^{-1}. \end{cases}$$

Remark. It is clear that the Green function G_τ depends on τ. When $\tau = 0$, (1.1.1) is reduced to

$$\sum_{i=1}^{m-2} \alpha_i - 1 + \exp(-\gamma) - \sum_{i=1}^{m-2} \alpha_i \exp(-\gamma(\eta_i)) \neq 0 \tag{1.1.2}$$

where m is an integer number > 3, $0 < \eta_1 < \eta_2 < \cdots < \eta_{m-2} < 1$, $\alpha_i \in \mathbf{R}$ $(i = 1, 2, \ldots, m-2)$. Then the m-point boundary problem

$$\begin{cases} \ddot{u}_{x,f}(t) + \gamma \dot{u}_{x,f}(t) = f(t), \ t \in [0, 1] \\ u_{x,f}(0) = x, \ u_{x,f}(1) = \sum_{i=1}^{m-2} \alpha_i u_{x,f}(\eta_i) \end{cases}$$

has a unique $C_E^2([0,1])$-solution (resp. $W_E^{2,1}([0,1])$-solution), $u_{x,f}$, with integral representation formulas

$$\begin{cases} u_{x,f}(t) = e_x(t) + \int_0^1 G_0(t,s)f(s)ds, \ t \in [0,1] \\ \dot{u}_{x,f}(t) = \dot{e}_x(t) + \int_0^1 \frac{\partial G_0}{\partial t}(t,s)f(s)ds, \ t \in [0,1] \end{cases}$$

where

$$\begin{cases} e_x(t) = x + A_0(1 - \sum_{i=1}^{m-2} \alpha_i)(1 - \exp(-\gamma t))x, \\ \dot{e}_x(t) = \gamma A_0 \left(1 - \sum_{i=1}^{m-2} \alpha_i\right) \exp(-\gamma t)x, \\ A_0 = \left(\sum_{i=1}^{m-2} \alpha_i - 1 + \exp(-\gamma) - \sum_{i=1}^{m-2} \alpha_i \exp(-\gamma(\eta_i))\right)^{-1}. \end{cases}$$

This remark and its notation will be used in the next section.

3. Existence of Optimal Controls

Let us recall the following denseness result based on Lyapunov theorem. See e.g. [12, 28].

Proposition 3.1. *Let E be a separable Banach space. Let $\Gamma : [0,T] \to cwk(E)$ be a convex weakly compact valued measurable and integrably bounded mapping. Let $ext(\Gamma) : t \mapsto ext(\Gamma(t))$ where $ext(\Gamma(t))$ is the set of extreme points of $\Gamma(t)(t \in [0,T])$. Then the set S_Γ^1 of all integrable selections of Γ is convex and $\sigma(L_E^1, L_{E^*}^\infty)$-compact and the set of all integrable selections $S_{ext(\Gamma)}^1$ of $ext(\Gamma)$ is dense in S_Γ^1 with respect to this topology.*

Proof. See e.g. [12, 28]. □

In this section we will assume that the hypotheses and notations of Lemma 2.1 hold with $\tau = 0$.

Theorem 3.1. *With the hypotheses and notations of Proposition 3.1, let E be a separable Banach space and let $\Gamma : [0,T] \to ck(E)$ be a convex compact valued measurable and integrably bounded mapping. Let us following (SODE)*

$$(SODE)_\Gamma \begin{cases} \ddot{u}_f(t) + \gamma \dot{u}_f(t) = f(t), \ f \in S^1_\Gamma \\ u_f(0) = x, \quad u_f(1) = \sum_{i=1}^{m-2} \alpha_i u_f(\eta_i) \end{cases}$$

$$(SODE)_{ext(\Gamma)} \begin{cases} \ddot{u}_g(t) + \gamma \dot{u}_g(t) = g(t), \ g \in S^1_{ext(\Gamma)} \\ u_g(0) = x, \quad u_g(1) = \sum_{i=1}^{m-2} \alpha_i u_g(\eta_i). \end{cases}$$

Then the set $\{u_f : f \in S^1_\Gamma\}$ of $W^{2,1}_E([0,1])$-solutions to $(SODE)_\Gamma$ is compact in $C^1_E([0,1])$ and the set $\{u_g : g \in S^1_{ext(\Gamma)}\}$ of $W^{2,1}_E([0,1])$-solutions to $(SODE)_{ext(\Gamma)}$ is dense in the compact set $\{u_f : f \in S^1_\Gamma\}$ of $W^{2,1}_E([0,1])$-solutions to $(SODE)_\Gamma$.

Proof. Step 1. Compactness of the solution set $\{u_f : f \in S^1_\Gamma\}$ in $C^1_E([0,1])$.

Let (u_{f_n}) be a sequence of $W^{2,1}_E([0,1])$-solutions to $(SODE)_\Gamma$. As S^1_Γ is $\sigma(L^1_E, L^\infty_{E^*})$-compact, by Eberlein–Smulian theorem, we may assume that (f_n) $\sigma(L^1_E, L^\infty_{E^*})$-converges to $f_\infty \in S^1_\Gamma$. From the properties of the Green function G_0 in Lemma 2.1 (by taking $\tau = 0$) we have, for each $n \in \mathbf{N}$,

$$u_{f_n}(t) = e_x(t) + \int_0^1 G_0(t,s) f_n(s)ds, \ t \in [0,1], \tag{3.1.1}$$

$$\dot{u}_{f_n}(t) = \dot{e}_x(t) + \int_0^1 \frac{\partial G_0}{\partial t}(t,s) f_n(s)ds, \ t \in [0,1], \tag{3.1.2}$$

$$\ddot{u}_{f_n}(t) + \gamma \dot{u}_{f_n}(t) = f_n(t) \in \Gamma(t), a.e. \ t \in [0,1] \tag{3.1.3}$$

with

$$\begin{cases} e_x(t) = x + A_0(1 - \sum_{i=1}^{m-2} \alpha_i)(1 - \exp(-\gamma t))x, \ t \in [0,1] \\ \dot{e}_x(t) = \gamma A_0 \left(1 - \sum_{i=1}^{m-2} \alpha_i \right) \exp(-\gamma t)x, \ t \in [0,1] \\ A_0 = \left(\sum_{i=1}^{m-2} \alpha_i - 1 + \exp(-\gamma) - \sum_{i=1}^{m-2} \alpha_i \exp(-\gamma(\eta_i)) \right)^{-1}. \end{cases}$$

On the other hand, from definition of the Green function G_0 in Lemma 2.1(iv) and (3.1.1), it is not difficult to show that $\{u_{f_n} : n \in \mathbf{N}\}$ is equicontinuous in $C_E([0,1])$. Indeed, let $t, t' \in [0,1]$, from (3.1.1) and (iv), we have the estimate

$$\begin{aligned}
&||u_{f_n}(t) - u_{f_n}(t')|| \\
&\leq ||e_x(t) - e_x(t')|| + \int_0^1 |G_0(t,s) - G_0(t',s)|\, ||\ddot{u}_{f_n}(s) + \gamma \dot{u}_{f_n}(s)||ds \\
&\leq ||e_x(t) - e_x(t')|| + \int_0^1 |G_0(t,s) - G_0(t',s)|\, |\Gamma(s)|ds.
\end{aligned}$$

Further, for each $t \in [0,1]$ $\{u_{f_n}(t) : n \in \mathbf{N}\}$ is relatively compact because it is included in the norm compact set $e_x(t) + \int_0^1 G_0(t,s)\Gamma(s)ds$ (see e.g. [12, 14]). So by Ascoli's theorem, $\{u_{f_n} : n \in \mathbf{N}\}$ is relatively compact in $C_E([0,1])$. Similarly using the properties of $\frac{\partial G_0}{\partial t}$ in Lemma 2.1 and (3.1.2) we deduce that $\{\dot{u}_{f_n} : n \in \mathbf{N}\}$ is equicontinuous in $C_E([0,1])$. In addition, the set $\{\dot{u}_{f_n}(t) : n \in \mathbf{N}\}$ is included in the compact set $\dot{e}_x(t) + \int_0^1 \frac{\partial G_0}{\partial t}(t,s)\,\Gamma(s)\,ds$. So $\{\dot{u}_{f_n} : n \in \mathbf{N}\}$ is relatively compact in $C_E([0,1])$ by Ascoli's theorem. From the above facts, we deduce that there exists a subsequence of $(u_{f_n})_{n\in\mathbf{N}}$ still denoted by $(u_{f_n})_{n\in\mathbf{N}}$ which converges uniformly to $u^\infty \in C_E([0,1])$ with $u^\infty(0) = x$, $u^\infty(1) = \sum_{i=1}^{m-2} \alpha_i u^\infty(\eta_i)$. Similarly, we may assume that $(\dot{u}_{f_n})$ converges uniformly to $v^\infty \in C_E([0,1])$. Furthermore, by the above facts, it is easy to see that $(\ddot{u}_{f_n})$ $\sigma(L^1_E, L^\infty_{E^*})$-converges to $w^\infty \in L^1_E([0,1])$. For every $t \in [0,1]$, using the representation formula (3.1.1), we have

$$\begin{aligned}
u^\infty(t) &= \lim_{n\to\infty} u_{f_n}(t) = e_x(t) + \lim_{n\to\infty} \int_0^1 G_0(t,s)(\ddot{u}_{f_n}(s) + \gamma \dot{u}_{f_n}(s))ds \\
&= e_x(t) + \lim_{n\to\infty} \int_0^1 G_0(t,s)\ddot{u}_{f_n}(s)ds + \gamma \lim_{n\to\infty} \int_0^1 G_0(t,s)\dot{u}_{f_n}(s)ds \\
&= e_x(t) + \int_0^1 G_0(t,s)w^\infty(s)ds + \gamma \int_0^1 G_0(t,s)v^\infty(s)ds \\
&= e_x(t) + \int_0^1 G_0(t,s)(w^\infty(s) + \gamma v^\infty(s))ds. \qquad (3.1.4)
\end{aligned}$$

From (3.1.4) and Lemma 2.1(iv), we deduce that u^∞ is derivable and its derivative $\dot{u}^\infty$ is given by

$$\dot{u}^\infty(t) = \dot{e}_x(t) + \int_0^1 \frac{\partial G_0}{\partial t}(t,s)(w^\infty(s) + \gamma v^\infty(s))ds, \forall t \in [0,1]. \quad (3.1.5)$$

Now using the integral representation formula (3.1.2) we have, for every $t \in [0, 1]$,

$$
\begin{aligned}
v^{\infty}(t) &= \lim_{n\to\infty} \dot{u}_{f_n}(t) = \dot{e}_x(t) + \lim_{n\to\infty} \int_0^1 \frac{\partial G_0}{\partial t}(t,s)(\ddot{u}_{f_n}(s) + \gamma \dot{u}_{f_n}(s))ds \\
&= \dot{e}_x(t) + \lim_{n\to\infty} \int_0^1 \frac{\partial G_0}{\partial t}(t,s)\ddot{u}_{f_n}(s)ds \\
&\quad + \gamma \lim_{n\to\infty} \int_0^1 \frac{\partial G_0}{\partial t}(t,s)\dot{u}_{f_n}(s)ds \\
&= \dot{e}_x(t) + \int_0^1 \frac{\partial G_0}{\partial t}(t,s)w^{\infty}(s)ds + \gamma \int_0^1 \frac{\partial G_0}{\partial t}(t,s)v_{\infty}(s)ds \\
&= \dot{e}_x(t) + \int_0^1 \frac{\partial G_0}{\partial t}(t,s)(w^{\infty}(s) + \gamma v^{\infty}(s))ds \qquad (3.1.6)
\end{aligned}
$$

so that by (3.1.5) and (3.1.6) we get $v^{\infty} = \dot{u}^{\infty}$. Now invoking Lemma 2.1(v) and using (3.1.4) we get

$$
\ddot{u}^{\infty}(t) + \gamma \dot{u}^{\infty}(t) = w^{\infty}(t) + \gamma v^{\infty}(t) = w^{\infty}(t) + \gamma \dot{u}^{\infty}(t) \quad a.e. \quad t \in [0, 1].
$$

Thus we get $\ddot{u}^{\infty}(t) = w^{\infty}(t)\, a.e.\, t \in [0, 1]$ so that by (3.1.4)

$$
\begin{cases}
u^{\infty}(t) = e_x(t) + \displaystyle\int_0^1 G_0(t,s)(\ddot{u}^{\infty}(s) + \gamma \dot{u}^{\infty}(s))ds, \quad t \in [0, 1] \\
u^{\infty}(0) = x, \quad u^{\infty}(1) = \displaystyle\sum_{i=1}^{m-2} \alpha_i u^{\infty}(\eta_i).
\end{cases} \qquad (3.1.7)
$$

Step 2. Main fact: u^{∞} coincides with the $W_E^{2,1}([0, 1])$-solution u_{f_∞} associated with $f_\infty \in S_\Gamma^1$ to

$$
\begin{cases}
\ddot{u}_{f_\infty}(t) + \gamma \dot{u}_{f_\infty}(t) = f_\infty(t), \\
u_{f_\infty}(0) = x, \quad u_{f_\infty}(1) = \displaystyle\sum_{i=1}^{m-2} \alpha_i u_{f_\infty}(\eta_i).
\end{cases} \qquad (3.1.8)
$$

Remember that

$$
\begin{cases}
\ddot{u}_{f_n}(t) + \gamma \dot{u}_{f_n}(t) = f_n(t), \\
u_{f_n}(0) = x, \quad u_{f_n}(1) = \displaystyle\sum_{i=1}^{m-2} \alpha_i u_{f_n}(\eta_i)
\end{cases}
$$

and by the above fact, $(\ddot{u}_{f_n} + \gamma\dot{u}_{f_n})$ converges weakly in $L^1_E([0,1])$ to $\ddot{u}^\infty + \gamma\dot{u}^\infty$. Let $v \in L^\infty_{E^*}([0,1])$. Multiply scalarly the equation

$$\ddot{u}_{f_n}(t) + \gamma\dot{u}_{f_n}(t) = f_n(t)$$

by $v(t)$ and integrating on $[0,1]$ yields

$$\int_0^1 \langle v(t), \ddot{u}_{f_n}(t) + \gamma\dot{u}_{f_n}(t)\rangle dt = \int_0^1 \langle v(t), f_n(t)\rangle dt. \tag{3.1.9}$$

It is clear that

$$\lim_{n\to\infty}\int_0^1 \langle v(t), \ddot{u}_{f_n}(t) + \gamma\dot{u}_{f_n}(t)\rangle dt = \int_0^1 \langle v(t), \ddot{u}^\infty(t) + \gamma\dot{u}^\infty(t)\rangle dt$$
$$= \lim_{n\to\infty}\int_0^1 \langle v(t), f_n(t)\rangle dt = \int_0^1 \langle v(t), f_\infty(t)\rangle dt$$

so that

$$\ddot{u}^\infty + \gamma\dot{u}^\infty = f_\infty. \tag{3.1.10}$$

Using (3.1.7), (3.1.8), and (3.1.10) and uniqueness of solutions we get $u^\infty = u_{f_\infty}$. This proves the first part of the theorem, while the second part follows from Proposition 3.1 and the integral representation formulas. □

Now comes a direct application to the existence of optimal controls for the problem

$$\begin{cases} \ddot{u}_f(t) + \gamma\dot{u}_f(t) = f(t),\ f \in S^1_\Gamma \\ u_f(0) = x, \quad u_f(1) = \sum_{i=1}^{m-2} \alpha_i u_f(\eta_i), \end{cases} \tag{*}$$

$$\inf_{f\in S^1_\Gamma}\int_0^1 J(t, u_f(t), \dot{u}_f(t), \ddot{u}_f(t))dt. \tag{**}$$

Theorem 3.2. *Under the hypotheses and notations of Theorem 3.1, problem (*)–(**) admits an optimal control.*

Proof. Let us set $m := \inf_{f\in S^1_\Gamma}\int_0^1 J(t, u_f(t), \dot{u}_f(t), \ddot{u}_f(t))dt$. Let us consider a minimizing sequence $(u_{f_n}, \dot{u}_{f_n}, \ddot{u}_{f_n})$, that is

$$\lim_{n\to\infty}\int_0^1 J(t, u_{f_n}(t), \dot{u}_{f_n}(t), \ddot{u}_{f_n}(t))dt = m.$$

Since (f_n) is relatively weakly compact in $L^1_E([0,1])$, we may assume that (f_n) converges weakly in $L^1_E([0,1])$ to $\overline{f}$. Applying the arguments in the proof of Theorem 3.1 shows that (u_{f_n}) converges uniformly to $(u_{\overline{f}})$, $(\dot{u}_{f_n})$ converges uniformly to $\dot{u}_{\overline{f}}$ and $(\ddot{u}_{f_n})$ $\sigma(L^1_E, L^\infty_{E^*})$ -converges to $\ddot{u}_{\overline{f}}$ with

$$\ddot{u}_{\overline{f}}(t) + \gamma \dot{u}_{\overline{f}}(t) = \overline{f}(t),$$

$$u_{\overline{f}}(0) = x, \quad u_{\overline{f}}(1) = \sum_{i=1}^{m-2} \alpha_i u_{\overline{f}}(\eta_i).$$

Now apply the lower semicontinuity for integral functionals ([14], Theorem 8.1.6) yields

$$\liminf_{n\to\infty} \int_0^1 J(t, u_{f_n}(t), \dot{u}_{f_n}(t), \ddot{u}_{f_n}(t))dt \geq \int_0^1 J(t, u_{\overline{f}}(t), \dot{u}_{\overline{f}}(t), \ddot{u}_{\overline{f}}(t))dt \geq m.$$

Hence we conclude that

$$m = \inf_{f \in S^1_\Gamma} \int_0^1 J(t, u_f(t), \dot{u}_f(t), \ddot{u}_f(t))dt = \int_0^1 J(t, u_{\overline{f}}(t), \dot{u}_{\overline{f}}(t), \ddot{u}_{\overline{f}}(t))dt.$$

□

Now along the paper we will assume that the hypotheses and notations of Lemma 2.1 hold.

4. Viscosity Property of the Value Function

The results given in Sect. 3 lead naturally to the problem of viscosity for the value function associated with a second order differential inclusion. Similar results dealing with ordinary differential equation (ODE) and evolution inclusion with control measures are available in [2, 7, 14, 16]. In this section we treat a new problem of value function in the context of second order ordinary differential equations (SODE) with m-point boundary condition. Assume that E is a separable Banach space, Z is a convex compact subset of E and S^1_Z is the set of all Lebesgue measurable mappings $f : [0, 1] \to Z$ (alias measurable selections of the constant mapping Z). For each $f \in S^1_Z$, let us denote by $u_{\tau,x,f}$ the trajectory solution associated with the control $f \in S^1_Z$ starting from x at time $\tau \in [0, \eta_1[$ to

$$(SODE) \begin{cases} \ddot{u}_{\tau,x,f}(t) + \gamma \dot{u}_{\tau,x,f}(t) = f(t), \ t \in [\tau, 1] \\ u_{\tau,x,f}(\tau) = x, \ u_{\tau,x,f}(1) = \sum_{i=1}^{m-2} \alpha_i u_{\tau,x,f}(\eta_i) \end{cases}$$

with the integral representation formulas

$$\begin{cases} u_{\tau,x,f}(t) = e_{\tau,x}(t) + \int_\tau^1 G_\tau(t,s) f(s) ds, \ t \in [\tau, 1] \\ \dot{u}_{\tau,x,f}(t) = \dot{e}_{\tau,x}(t) + \int_\tau^1 \frac{\partial G_\tau}{\partial t}(t,s) f(s) ds, \ t \in [\tau, 1] \end{cases} \tag{4.1}$$

and

$$\begin{cases} e_{\tau,x}(t) = x + A_\tau (1 - \sum_{i=1}^{m-2} \alpha_i)(1 - \exp(-\gamma(t-\tau)))x, \ t \in [\tau, 1] \\ \dot{e}_{\tau,x}(t) = \gamma A_\tau \left(1 - \sum_{i=1}^{m-2} \alpha_i\right) \exp(-\gamma(t-\tau))x, \ t \in [\tau, 1] \\ A_\tau = \left(\sum_{i=1}^{m-2} \alpha_i - \dot{1} + \exp(-\gamma(1-\tau)) - \sum_{i=1}^{m-2} \alpha_i \exp(-\gamma(\eta_i - \tau))\right)^{-1} \end{cases} \tag{4.2}$$

where the coefficient A_τ and the Green function G_τ are given in Lemma 2.1.

By the above considerations and Lemma 2.1(ii), it is easy to check that $\dot{u}_{\tau,x,f}$ are uniformly majorized by a continuous function $c_\tau : [\tau, 1] \to \mathbf{R}^+$, namely

$$\begin{aligned} ||\dot{u}_{\tau,x,f}(t)|| &\leq ||e_{\tau,x}(t)|| + \int_\tau^1 |\frac{\partial G_\tau}{\partial t}(t,s)| \, ||f(s)|| ds \\ &\leq ||e_{\tau,x}(t)|| + \int_\tau^1 |\frac{\partial G_\tau}{\partial t}(t,s)| |Z| ds = c_\tau(t), \ \forall t \in [\tau, 1]. \end{aligned} \tag{4.3}$$

It is worth mentioning that integral representation formulas (4.1) and (4.2) will be useful in the study of the value function we present below. Let us mention a useful lemma that is borrowed from ([16], Lemma 6.3) and ([7], Lemma 3.1).

Lemma 4.1. *Assume that (1.1.1) is satisfied. Let* $(t_0, x_0) \in [0, \eta_1[\times E$ *and let* Z *be a convex compact subset in* E. *Let* $\Lambda : [0, T] \times E \times Z \to \mathbf{R}$ *be an upper semicontinuous function such that the restriction of* Λ *to* $[0, T] \times B \times Z$ *is bounded on any bounded subset* B *of* E. *If*

$$max_{z \in Z} \Lambda(t_0, x_0, z) < -\eta < 0$$

for some $\eta > 0$, *then there exists* $\sigma > 0$ *such that*

$$\sup_{f \in S_Z^1} \left\{ \int_{t_0}^{t_0+\sigma} \Lambda(t, u_{t_0,x_0,f}(t), f(t)) dt \right\} < -\frac{\sigma\eta}{2}$$

where $u_{t_0,x_0,f}$ is the trajectory solution associated with the control $f \in S^1_Z$ starting from x_0 at time t_0 to

$$(SODE)\begin{cases} \ddot{u}_{t_0,x_0,f}(t) + \gamma \dot{u}_{t_0,x_0,f}(t) = f(t),\ t \in [t_0, 1] \\ u_{t_0,x_0,f}(t_0) = x_0,\ u_{t_0,x_0,f}(1) = \sum_{i=1}^{m-2} \alpha_i u_{t_0,x_0,f}(\eta_i). \end{cases}$$

Proof. By hypothesis, one has $\max_{z\in Z} \Lambda(t_0, x_0, z) < -\eta < 0$. As Λ is upper semi continuous, so is the function

$$(t,x) \mapsto \max_{z\in Z} \Lambda(t,x,z).$$

Hence there is $\varepsilon > 0$ such that

$$\max_{z\in Z} \Lambda(t,x,z) < -\frac{\eta}{2}$$

for $0 \le t - t_0 \le \varepsilon$ and $||x - x_0|| \le \varepsilon$. As $\dot{u}_{t_0,x_0,f}$ is uniformly bounded for all $f \in S^1_Z$ and for all $t \in [t_0, 1]$ by using the estimate (4.3) we can take $\sigma > 0$ such that $||u_{t_0,x_0,f}(t) - u_{t_0,x_0,f}(t_0)|| \le \varepsilon$ for all $t \in [t_0, t_0 + \sigma]$ and for all $f \in S^1_Z$. Then by integrating

$$\int_{t_0}^{t_0+\sigma} \Lambda(t, u_{t_0,x_0,f}(t), f(t))dt \le \int_{t_0}^{t_0+\sigma} [\max_{z\in Z} \Lambda(t, u_{t_0,x_0,f}(t), z)]dt < -\frac{\sigma\eta}{2}$$

for all $f \in S^1_Z$ and the result follows. □

For simplicity we deal first with a dynamic programming principle (DPP) for a value function V_J related to a bounded continuous function $J : [0,1] \times E \times Z \to \mathbf{R}$ associated with

$$(SODE)\begin{cases} \ddot{u}(t) + \gamma \dot{u}(t) = f(t),\ f \in S^1_Z,\ t \in [\tau, 1] \\ u(\tau) = x,\ u(1) = \sum_{i=1}^{m-2} \alpha_i u(\eta_i). \end{cases}$$

The following result is of importance in the statement of viscosity.

Theorem 4.1 (of Dynamic Programming Principle). *Let* (1.1.1) *holds. Let $x \in E$, $0 \le \tau < \eta_1 < .. < \eta_{m-2} < 1$ and $\sigma > 0$ such that $\tau + \sigma < \eta_1$. Assume that $J : [0,1] \times E \times E \to \mathbf{R}$ is bounded continuous such that $J(t,x,.)$ is convex on E for every $(t,x) \in [0,1] \times E$. Let us consider the value function*

$$V_J(\tau, x) = \sup_{f\in S^1_Z} \{\int_\tau^1 J(t, u_{\tau,x,f}(t), f(t))dt\}, \quad (\tau, x) \in [0, \eta_1[\times E$$

where $u_{\tau,x,f}$ is the trajectory solution on $[\tau, 1]$ associated the control $f \in S^1_Z$ starting from x at time τ to

$$(SODE)\begin{cases} \ddot{u}_{\tau,x,f}(t) + \gamma \dot{u}_{\tau,x,f}(t) = f(t),\ t \in [\tau, 1] \\ u_{\tau,x,f}(\tau) = x,\ u_{\tau,x,f}(1) = \sum_{i=1}^{m-2} \alpha_i u_{\tau,x,f}(\eta_i). \end{cases} \tag{4.4}$$

Then the following hold

$$V_J(\tau, x) = \sup_{f \in S^1_Z} \left\{ \int_\tau^{\tau+\sigma} J(t, u_{\tau,x,f}(t), f(t))dt + V_J(\tau+\sigma, u_{\tau,x,f}(\tau+\sigma)) \right\}$$

with

$$V_J(\tau+\sigma, u_{\tau,x,f}(\tau+\sigma)) = \sup_{g \in S^1_Z} \left\{ \int_{\tau+\sigma}^1 J(t, v_{\tau+\sigma,u_{\tau,x,f}(\tau+\sigma),g}(t), g(t))dt \right\}$$

where $v_{\tau+\sigma,u_{\tau,x,f}(\tau+\sigma),g}$ denotes the trajectory solution on $[\tau+\sigma, 1]$ associated with the control $g \in S^1_Z$ starting from $u_{\tau,x,f}(\tau+\sigma)$ at time $\tau+\sigma$ to[1]

$$(SODE)\begin{cases} \ddot{v}_{\tau+\sigma,u_{\tau,x,f}(\tau+\sigma),g}(t) + \gamma \dot{v}_{\tau+\sigma,u_{\tau,x,f}(\tau+\sigma),g}(t) = g(t), \\ \quad t \in [\tau+\sigma, 1] \\ v_{\tau+\sigma,u_{\tau,x,f}(\tau+\sigma),g}(\tau+\sigma) = u_{\tau,x,f}(\tau+\sigma), \\ v_{\tau+\sigma,u_{\tau,x,f}(\tau+\sigma),g}(1) = \sum_{i=1}^{m-2} \alpha_i v_{\tau+\sigma,u_{\tau,x,f}(\tau+\sigma),g}(\eta_i). \end{cases} \tag{4.5}$$

Proof. Let

$$W_J(\tau, x) := \sup_{f \in S^1_Z} \left\{ \int_\tau^{\tau+\sigma} J(t, u_{\tau,x,f}(t), f(t))dt + V_J(\tau+\sigma, u_{\tau,x,f}(\tau+\sigma)) \right\}.$$

For any $f \in S^1_Z$, we have

$$\int_\tau^1 J(t, u_{\tau,x,f}(t), f(t))dt$$
$$= \int_\tau^{\tau+\sigma} J(t, u_{\tau,x,f}(t), f(t))dt + \int_{\tau+\sigma}^1 J(t, u_{\tau,x,f}(t), f(t))dt.$$

[1] It is necessary to write completely the expression of the trajectory $v_{\tau+\sigma,u_{\tau,x,f}(\tau+\sigma),g}$ that depends on $(f, g) \in S^1_Z \times S^1_Z$ in order to get the lower semi-continuous dependence with respect to $f \in S^1_Z$ of $V_J(\tau+\sigma, u_{\tau,x,f}(\tau+\sigma))$.

By the definition of $V_J(\tau+\sigma, u_{\tau,x,f}(\tau+\sigma))$ we have

$$V_J(\tau+\sigma, u_{\tau,x,f}(\tau+\sigma)) \geq \int_{\tau+\sigma}^{1} J(t, u_{\tau,x,f}(t), f(t))dt.$$

It follows that

$$\begin{aligned}&\int_{\tau}^{1} J(t, u_{\tau,x,f}(t), f(t))dt \\ &\leq \int_{\tau}^{\tau+\sigma} J(t, u_{\tau,x,f}(t), f(t))dt + V_J(\tau+\sigma, u_{\tau,x,f}(\tau+\sigma)).\end{aligned}$$

By taking the supremum on S^1_Z in this inequality we get

$$\begin{aligned}V_J(\tau, x) &\leq \sup_{f\in S^1_Z} \{\int_{\tau}^{\tau+\sigma} J(t, u_{\tau,x,f}(t), f(t))dt + V_J(\tau+\sigma, u_{\tau,x,f}(\tau+\sigma))\} \\ &= W_J(\tau, x).\end{aligned}$$

Let us prove the converse inequality.

Main Fact: $f \to V_J(\tau+\sigma, u_{\tau,x,f}(\tau+\sigma))$ is lower semicontinuous on S^1_Z (endowed with the $\sigma(L^1_E, L^\infty_{E^*})$-topology).

Let us focus on the expression of $V_J(\tau+\sigma, u_{\tau,x,f}(\tau+\sigma))$

$$V_J(\tau+\sigma, u_{\tau,x,f}(\tau+\sigma)) = \sup_{g\in S^1_Z} \left\{ \int_{\tau+\sigma}^{1} J(t, v_{\tau+\sigma, u_{\tau,x,f}(\tau+\sigma), g}(t), g(t))dt \right\}$$

where $v_{\tau+\sigma, u_{\tau,x,f}(\tau+\sigma), g}$ denotes the trajectory solution on $[\tau+\sigma, 1]$ associated with the control $g \in S^1_Z$ starting from $u_{\tau,x,f}(\tau+\sigma)$ at time $\tau+\sigma$ to (SODE) (4.5). By the integral representation formulas (4.1) (4.2) given above we have

$$v_{\tau+\sigma, u_{\tau,x,f}(\tau+\sigma), g}(t) = e_{\tau+\sigma, u_{\tau,x,f}(\tau+\sigma)}(t) + \int_{\tau+\sigma}^{1} G_{\tau+\sigma}(t, s)g(s)ds$$

with

$$\begin{aligned}&e_{\tau+\sigma, u_{\tau,x,f}(\tau+\sigma)}(t) \\ &= u_{\tau,x,f}(\tau+\sigma) + A_{\tau+\sigma}(1 - \sum_{i=1}^{m-2} \alpha_i)(1 - \exp(-\gamma(t-(\tau+\sigma)))u_{\tau,x,f}(\tau+\sigma).\end{aligned}$$

It is already seen in the proof of Step 1 of Theorem 3.1 that $f \mapsto u_{\tau,x,f}$ from S^1_Z into $C_E([\tau, 1])$ is continuous when S^1_Z is endowed with the $\sigma(L^1_E, L^\infty_{E^*})$

topology and $C_E([\tau, 1])$ is endowed with the norm of uniform convergence, namely, when f_n $\sigma(L^1_E, L^\infty_{E^*})$-converges to $f \in S^1_Z$, then u_{τ,x,f_n} converges uniformly to $u_{\tau,x,f}$, this entails that

$$e_{\tau+\sigma,u_{\tau,x,f_n}}(\tau+\sigma)(t) \to e_{\tau+\sigma,u_{\tau,x,f}}(\tau+\sigma)(t)$$

for every $t \in [\tau, 1]$. Further, when g_n $\sigma(L^1_E, L^\infty_{E^*})$-converges to $g \in S^1_Z$, by compactness of Z, and the boundedness property of $G_{\tau+\sigma}(t,s)$ in Lemma 2.1, it is not difficult to check that

$$\int_{\tau+\sigma}^1 G_{\tau+\sigma}(t,s)g_n(s)ds \to \int_{\tau+\sigma}^1 G_{\tau+\sigma}(t,s)g(s)ds$$

for every $t \in [\tau, 1]$. Therefore

$$v_{\tau+\sigma,u_{\tau,x,f_n}(\tau+\sigma),g_n}(t) \to v_{\tau+\sigma,u_{\tau,x,f}(\tau+\sigma),g}(t)$$

for every $t \in [\tau, 1]$. Hence in view of ([14], Theorem 8.1.6) we deduce that

$$(f,g) \mapsto \int_{\tau+\sigma}^1 J(t, v_{\tau+\sigma,u_{\tau,x,f}(\tau+\sigma),g}(t), g(t))dt$$

is lower semicontinuous on $S^1_Z \times S^1_Z$ using the above fact and the convexity assumption on the integrand $J(t,x,.)$. Consequently $f \to V_J(\tau+\sigma, u_{\tau,x,f}(\tau+\sigma))$ is lower semicontinuous on S^1_Z. Hence the mapping

$$f \mapsto \int_\tau^{\tau+\sigma} J(t, u_{\tau,x,f}(t), f(t))dt + V_J(\tau+\sigma, u_{\tau,x,f}(\tau+\sigma))$$

is lower semicontinuous on S^1_Z. Since S^1_Z is weakly compact in $L^1_E([0,1])$, there is $f^1 \in S^1_Z$ such that

$$\begin{aligned} W_J(\tau,x) &= \sup_{f\in S^1_Z}\left\{\int_\tau^{\tau+\sigma} J(t, u_{\tau,x,f}(t), f(t))dt + V_J(\tau+\sigma, u_{\tau,x,f}(\tau+\sigma))\right\} \\ &= \int_\tau^{\tau+\sigma} J(t, u_{\tau,x,f^1}(t), f^1(t))dt + V_J(\tau+\sigma, u_{\tau,x,f^1}(\tau+\sigma)). \end{aligned}$$

Similarly there is $g^2 \in S^1_Z$ such that

$$\begin{aligned} V_J(\tau+\sigma, u_{\tau,x,f^1}(\tau+\sigma)) &= \sup_{g\in S^1_Z}\left\{\int_{\tau+\sigma}^1 J(t, v_{\tau+\sigma,u_{\tau,x,f^1}(\tau+\sigma),g}(t), g(t))dt\right\} \\ &= \int_{\tau+\sigma}^1 J(t, v_{\tau+\sigma,u_{\tau,x,f^1}(\tau+\sigma),g^2}(t), g^2(t))dt \end{aligned}$$

where $v_{\tau+\sigma,u_{\tau,x,f^1}(\tau+\sigma),g^2}(t)$ denotes the trajectory solution on $[\tau+\sigma,1]$ associated with the control $g^2 \in S^1_Z$ starting from $u_{\tau,x,f^1}(\tau+\sigma)$ at time $\tau+\sigma$ to

$$(SODE)\begin{cases} \ddot{v}_{\tau+\sigma,u_{\tau,x,f^1}(\tau+\sigma),g^2}(t)+\gamma\dot{v}_{\tau+\sigma,u_{\tau,x,f^1}(\tau+\sigma),g^2}(t)=g^2(t), \\ \quad t\in[\tau+\sigma,1] \\ v_{\tau+\sigma,u_{\tau,x,f^1}(\tau+\sigma),g^2}(\tau+\sigma)=u_{\tau,x,f^1}(\tau+\sigma), \\ v_{\tau+\sigma,u_{\tau,x,f^1}(\tau+\sigma),g^2}(1)=\sum_{i=1}^{m-2}\alpha_i v_{\tau+\sigma,u_{\tau,x,f^1}(\tau+\sigma),g^2}(\eta_i). \end{cases}$$

Let us set

$$\overline{f}:=1_{[\tau,\tau+\sigma]}f^1+1_{[\tau+\sigma,1]}f^2.$$

Then $\overline{f}\in S^1_Z$ (because S^1_Z is decomposable). Let $w_{\tau,x,\overline{f}}$ be the trajectory solution on $[\tau,1]$ associated with $\overline{f}\in S^1_Z$, that is

$$\ddot{w}_{\tau,x,\overline{f}}(t)+\gamma\ddot{w}_{\tau,x,\overline{f}}(t)=\overline{f}(t),\ t\in[\tau,1],$$

$$w_{\tau,x,\overline{f}}(\tau)=x,\ w_{\tau,x,\overline{f}}(1)=\sum_{i=1}^{m-2}\alpha_i w_{\tau,x,\overline{f}}(\eta_i).$$

By uniqueness of solution we have

$$w_{\tau,x,\overline{f}}(t)=u_{\tau,x,f^1}(t),\ \forall t\in[\tau,\tau+\sigma],$$
$$w_{\tau,x,\overline{f}}(t)=v_{\tau+\sigma,u_{\tau,x,f^1}(\tau+\sigma),g^2}(t),\ \forall t\in[\tau+\sigma,1].$$

Coming back to the expression of V_J and W_J we have

$$\begin{aligned} &W_J(\tau,x)\\ &=\int_\tau^{\tau+\sigma}J(t,u_{\tau,x,f^1}(t),f^1(t))dt+\int_{\tau+\sigma}^1 J(t,v_{\tau+\sigma,u_{\tau,x,f^1}(\tau+\sigma),g^2}(t),g^2(t))dt\\ &=\int_\tau^1 J(t,w_{\tau,x,\overline{f}}(t),\overline{f}(t))dt\\ &\le \sup_{f\in S^1_Z}\{\int_\tau^1 J(t,u_{\tau,x,f}(t),f(t))dt\}=V_J(\tau,x). \end{aligned}$$

□

Here are our results on viscosity of solutions for the value function.

Theorem 4.2 (of Viscosity Subsolution). *Assume that E is a separable Hilbert space. Assume (1.1.1) and $J:[0,1]\times E\times E\to\mathbf{R}$ is bounded* *continuous such that $J(t, x, .)$ is convex on E for every $(t, x) \in [0, 1] \times E$. Let us consider the value function*

$$V_J(\tau, x) = \sup_{f \in S^1_Z} \left\{ \int_\tau^1 J(t, u_{\tau,x,f}(t), f(t))dt \right\}, \ (\tau, x) \in [0, \eta_1[\times E$$

where $u_{\tau,x,f}$ is the trajectory solution on $[\tau, 1]$ associated the control $f \in S^1_Z$ starting from $x \in E$ at time τ to

$$(SODE) \begin{cases} \ddot{u}_{\tau,x,f}(t) + \gamma \dot{u}_{\tau,x,f}(t) = f(t), \ t \in [\tau, 1] \\ u_{\tau,x,f}(\tau) = x, \ u_{\tau,x,f}(1) = \sum_{i=1}^{m-2} \alpha_i u_{\tau,x,f}(\eta_i). \end{cases}$$

Then V_J satisfies a viscosity property: For any $\varphi \in C^1([0, 1] \times E)$ such that V_J reaches a local maximum at $(t_0, x_0) \in [0, \eta_1[\times E$, then

$$\frac{\partial \varphi}{\partial t}(t_0, x_0) + \max_{z \in Z}\{J(t_0, x_0, z)\} + \delta^*(\nabla\varphi(t_0, x_0), \dot{e}_{t_0,x_0}(t_0) + \int_{t_0}^1 \frac{\partial G_{t_0}}{\partial t}(t_0, s)Zds) \geq 0.$$

Proof. Assume by contradiction that there exist a $\varphi \in C^1([0, 1] \times E)$ such that V_J reaches a local maximum at $(t_0, x_0) \in [0, \eta_1[\times E$ for which

$$\frac{\partial \varphi}{\partial t}(t_0, x_0) + \max_{z \in Z}\{J(t_0, x_0, z)\} + \delta^*(\nabla\varphi(t_0, x_0), \dot{e}_{t_0,x_0}(t_0) + \int_{t_0}^1 \frac{\partial G_{t_0}}{\partial t}(t_0, s)Zds)$$
$$\leq -\eta < 0$$

for some $\eta > 0$. Applying Lemma 4.1, by taking

$$\Lambda(t, x, z) = J(t, x, z) + \delta^*(\nabla\varphi(t, x), \dot{e}_{t_0,x_0}(t) + \int_{t_0}^1 \frac{\partial G_{t_0}}{\partial t}(t, s)Zds) + \frac{\partial \varphi}{\partial t}(t, x)$$

yields $\sigma > 0$ such that

$$\sup_{f \in S^1_Z} \left\{ \int_{t_0}^{t_0+\sigma} J(t, u_{t_0,x_0,f}(t), f(t))dt \right.$$
$$+ \int_{t_0}^{t_0+\sigma} \delta^*(\nabla\varphi(t, u_{t_0,x_0,f}(t)), \dot{e}_{t_0,x_0}(t) + \int_{t_0}^1 \frac{\partial G_{t_0}}{\partial t}(t, s)Zds)dt$$
$$\left. + \int_{t_0}^{t_0+\sigma} \frac{\partial \varphi}{\partial t}(t, u_{t_0,x_0,f}(t))dt \right\}$$
$$< -\frac{\sigma\eta}{2} \tag{4.2.1}$$

where $u_{t_0,x_0,f}$ is the trajectory solution associated with the control $f \in S^1_Z$ starting from x_0 at time t_0 to

$$(SODE)\begin{cases} \ddot{u}_{t_0,x_0,f}(t) + \gamma \dot{u}_{t_0,x_0,f}(t) = f(t),\ t \in [t_0, 1] \\ u_{t_0,x_0,f}(t_0) = x_0,\ u_{t_0,x_0,f}(1) = \sum_{i=1}^{m-2} \alpha_i u_{t_0,x_0,f}(\eta_i). \end{cases}$$

Applying the dynamic programming principle (Theorem 4.1) gives

$$V_J(t_0, x_0) = \sup_{f \in S^1_Z} \left\{ \int_{t_0}^{t_0+\sigma} J(t, u_{t_0,x_0,f}(t), f(t))dt + V_J(t_0+\sigma, u_{t_0,x_0,f}(t_0+\sigma)) \right\}. \tag{4.2.2}$$

Since $V_J - \varphi$ has a local maximum at (t_0, x_0), for small enough σ

$$\begin{aligned} V_J(t_0, x_0) - \varphi(t_0, x_0) \geq V_J(t_0+\sigma, u_{t_0,x_0,f}(t_0+\sigma)) \\ -\varphi(t_0+\sigma, u_{t_0,x_0,f}(t_0+\sigma)) \end{aligned} \tag{4.2.3}$$

for all $f \in S^1_Z$. By (4.2.2), for each $n \in \mathbf{N}$, there is $f^n \in S^1_Z$ such that

$$\begin{aligned} V_J(t_0, x_0) \leq \int_{t_0}^{t_0+\sigma} J(t, u_{t_0,x_0,f^n}(t), f^n(t))dt + V_J(t_0 \\ + \sigma, u_{t_0,x_0,f^n}(t_0+\sigma)) + \frac{1}{n}. \end{aligned} \tag{4.2.4}$$

From (4.2.3) and (4.2.4) we deduce that

$$\begin{aligned} V_J(t_0+\sigma, u_{t_0,x_0,f^n}(t_0+\sigma)) - \varphi(t_0+\sigma, u_{t_0,x_0,f^n}(t_0+\sigma)) \\ \leq \int_{t_0}^{t_0+\sigma} J(t, u_{t_0,x_0,f^n}(t), f^n(t))dt + \frac{1}{n} \\ -\varphi(t_0, x_0) + V_J(t_0+\sigma, u_{t_0,x_0,f^n}(t_0+\sigma)). \end{aligned}$$

Therefore we have

$$\begin{aligned} 0 \leq \int_{t_0}^{t_0+\sigma} J(t, u_{t_0,x_0,f^n}(t), f^n(t))dt \\ + \varphi(t_0+\sigma, u_{t_0,x_0,f^n}(t_0+\sigma)) - \varphi(t_0, x_0) + \frac{1}{n}. \end{aligned} \tag{4.2.5}$$

As $\varphi \in C^1([0, 1] \times E)$

$$\begin{aligned} &\varphi(t_0+\sigma, u_{t_0,x_0,f^n}(t_0+\sigma)) - \varphi(t_0, x_0) \\ &= \int_{t_0}^{t_0+\sigma} \langle \nabla\varphi(t, u_{t_0,x_0,f^n}(t)), \dot{u}_{t_0,x_0,f^n}(t)\rangle dt + \int_{t_0}^{t_0+\sigma} \frac{\partial \varphi}{\partial t}(t, u_{t_0,x_0,f^n}(t))dt. \end{aligned} \tag{4.2.6}$$

Applying the integral representation formulas (4.1) and (4.2) gives

$$\begin{cases} u_{t_0,x_0,f^n}(t) = e_{t_0,x_0}(t) + \int_{t_0}^{1} G_{t_0}(t,s)f^n(s)ds,\ t \in [t_0,1] \\ \dot{u}_{t_0,x_0,f^n}(t) = \dot{e}_{t_0,x_0}(t) + \int_{t_0}^{1} \frac{\partial G_{t_0}}{\partial t}(t,s)f^n(s)ds,\ t \in [t_0,1] \end{cases}$$

with

$$\begin{cases} e_{t_0,x_0}(t) = x_0 + A_{t_0}(1 - \sum_{i=1}^{m-2}\alpha_i)(1 - \exp(-\gamma(t-t_0)))x_0,\ \forall t \in [t_0,1] \\ \dot{e}_{t_0,x_0}(t) = \gamma A_{t_0}\left(1 - \sum_{i=1}^{m-2}\alpha_i\right)\exp(-\gamma(t-t_0))x_0,\ \forall t \in [t_0,1] \\ A_{t_0} = \left(\sum_{i=1}^{m-2}\alpha_i - 1 + \exp(-\gamma(1-t_0)) - \sum_{i=1}^{m-2}\alpha_i\exp(-\gamma(\eta_i - t_0))\right)^{-1} \end{cases}$$

where the coefficient A_{t_0} and the Green function G_{t_0} are defined in Lemma 2.1. Then from (4.2.6) we get the estimation

$$\begin{aligned} &\varphi(t_0+\sigma, u_{t_0,x_0,f^n}(t_0+\sigma)) - \varphi(t_0,x_0) \\ &= \int_{t_0}^{t_0+\sigma} \langle \nabla\varphi(t, u_{t_0,x_0,f^n}(t)), \dot{e}_{t_0,x_0}(t) + \int_0^1 \frac{\partial G_{t_0}}{\partial t}(t,s)f^n(s)ds\rangle dt \\ &+ \int_{t_0}^{t_0+\sigma} \frac{\partial\varphi}{\partial t}(t, u_{t_0,x_0,f^n}(t))dt. \end{aligned} \tag{4.2.7}$$

Since $f^n(s) \in Z$ for all $s \in [t_0,1]$, it follows that

$$\frac{\partial G_{t_0}}{\partial t}(t,s)f^n(s) \in \frac{\partial G_{t_0}}{\partial t}(t,s)Z$$

for all $t, s \in [t_0,1]$. From (4.2.7) and this inclusion we get

$$\begin{aligned} &\varphi(t_0+\sigma, u_{t_0,x_0,f^n}(t_0+\sigma)) - \varphi(t_0,x_0) \\ &\le \int_{t_0}^{t_0+\sigma} \delta^*(\nabla\varphi(t, u_{t_0,x_0,f^n}(t)), \dot{e}_{t_0,x_0}(t) + \int_{t_0}^1 \frac{\partial G_{t_0}}{\partial t}(t,s)Zds)dt \\ &+ \int_{t_0}^{t_0+\sigma} \frac{\partial\varphi}{\partial t}(t, u_{t_0,x_0,f^n}(t))dt. \end{aligned} \tag{4.2.8}$$

Put the estimation (4.2.8) in (4.2.5) we get

$$\begin{aligned}0 \le &\int_{t_0}^{t_0+\sigma} J(t, u_{t_0,x_0,f^n}(t), f^n(t))dt\\ &+\int_{t_0}^{t_0+\sigma} \delta^*(\nabla\varphi(t, u_{t_0,x_0,f^n}(t)), \dot{e}_{t_0,x_0}(t)+\int_{t_0}^{1}\frac{\partial G_{t_0}}{\partial t}(t,s)Zds)dt\\ &+\int_{t_0}^{t_0+\sigma}\frac{\partial\varphi}{\partial t}(t, u_{t_0,x_0,f^n}(t))dt+\frac{1}{n}. \end{aligned} \tag{4.2.9}$$

By combining (4.2.1) and (4.2.9) we get the estimation

$$\begin{aligned}0 \le &\int_{t_0}^{t_0+\sigma} J(t, u_{t_0,x_0,f^n}(t), f^n(t))dt\\ &+\int_{t_0}^{t_0+\sigma} \delta^*(\nabla\varphi(t, u_{t_0,x_0,f^n}(t)), \dot{e}_{t_0,x_0}(t)+\int_{t_0}^{1}\frac{\partial G_{t_0}}{\partial t}(t,s)Zds)dt\\ &+\int_{t_0}^{t_0+\sigma}\frac{\partial\varphi}{\partial t}(t, u_{t_0,x_0,f^n}(t))dt+\frac{1}{n} < -\frac{\sigma\eta}{2}+\frac{1}{n}. \end{aligned} \tag{4.2.10}$$

Therefore we have that $0 < \frac{\sigma\eta}{2} < \frac{1}{n}$ for every $n \in \mathbf{N}$. Passing to the limit when n goes to ∞ in the preceding inequality yields a contradiction. □

5. Optimal Control Problem in Pettis Integration

We provide in this section some results in optimal control problems governed by an (SODE) with m-point boundary condition where the controls are Pettis-integrable. Here E is a separable Banach space. We recall and summarize some needed results on the Pettis integrability. Let $f : [0, 1] \to E$ be a scalarly integrable function, that is, for every $x^* \in E^*$, the scalar function $t \mapsto \langle x^*, f(t)\rangle$ is Lebesgue-integrable on $[0, 1]$. A scalarly integrable function $f : [0, 1] \to E$ is Pettis-integrable if, for every Lebesgue-measurable set A in $[0, 1]$, the weak integral $\int_A f(t)dt$ defined by $\langle x^*, \int_A f(t)dt\rangle = \int_A \langle x^*, f(t)\rangle\, dt$ for all $x^* \in E^*$ belongs to E. We denote by $P^1_E([0, 1], dt)$ the space of all Pettis-integrable functions $f : [0, 1] \to E$ endowed with the Pettis norm $||f||_{Pe} = \sup_{x^*\in\overline{B}_{E^*}} \int_0^1 |\langle x^*, f(t)\rangle|dt$. A mapping $f : [0, 1] \to E$ is Pettis-integrable iff the set $\{\langle x^*, f\rangle : ||x^*|| \le 1\}$ is uniformly integrable in the space $L^1_{\mathbf{R}}([0, 1], dt)$. More generally a convex compact valued mapping $\Gamma : [0, 1] \rightrightarrows E$ is scalarly integrable, if, for every $x^* \in E^*$, the scalar function $t \mapsto \delta^*(x^*, \Gamma(t))$ is Lebesgue-integrable on $[0, 1]$, Γ is Pettis-integrable if the set $\{\delta^*(x^*, \Gamma(.)) : ||x^*|| \le 1\}$ is uniformly integrable in the space $L^1_{\mathbf{R}}([0, 1], dt)$. In view of [[6], Theorem 4.2; or [14], Cor. 6.3.3] the set S_Γ^{Pe} of all Pettis-integrable selections of a convex compact valued Pettis-integrable mapping $\Gamma : [0,1] \rightrightarrows E$ is sequentially $\sigma(P_E^1, L^\infty \otimes E^*)$-compact. We refer to [19], for related results on the integration of Pettis-integrable multifunctions.

We provide some useful lemmas.

Lemma 5.1. *Let $G : [0,1] \times [0,1] \to \mathbf{R}$ be a mapping with the following properties*

(i) for each $t \in [0,1]$, $G(t,.)$ is Lebesgue-measurable on $[0,1]$,
(ii) for each $s \in [0,1]$, $G(.,s)$ is continuous on $[0,1]$,
(iii) there is a constant $M > 0$ such that $|G(t,s)| \le M$ for all $(t,s) \in [0,1] \times [0,1]$.

Let $f : [0,1] \to E$ be a Pettis-integrable mapping. Then the mapping

$$u_f : t \mapsto \int_0^1 G(t,s)f(s)ds$$

is continuous from $[0,1]$ into E, that is, $u_f \in C_E([0,1])$.

Proof. Let (t_n) be a sequence in $[0,1]$ such that $t_n \to t \in [0,1]$. Then we have the estimation

$$\sup_{x^* \in \overline{B}_{E^*}} |\langle x^*, \int_0^1 G(t_n,s)f(s)ds - \int_0^1 G(t,s)f(s)ds\rangle|$$
$$\le \sup_{x^* \in \overline{B}_{E^*}} \int_0^1 |G(t_n,s) - G(t,s)||\langle x^*, f(s)\rangle|ds.$$

As the sequence $(|G(t_n,.) - G(t,.)|)$ is bounded in $L_{\mathbf{R}}^\infty([0,1])$ and pointwise converges to 0, it converges to 0 uniformly on uniformly integrable subsets of $L_{\mathbf{R}}^1([0,1])$ in view of a lemma due to Grothendieck's [24], in others terms it converges to 0 with respect to the Mackey topology $\tau(L^\infty, L^1)$, see also [5] for a more general result concerning the Mackey topology for bounded sequences in $L_{E^*}^\infty$. Since the set $\{|\langle x^*, f(s)\rangle| : ||x^*|| \le 1\}$ is uniformly integrable in $L_{\mathbf{R}}^1([0,1])$, the second term in the above estimation goes to 0 when $t_n \to t$ showing that u_f is continuous on $[0,1]$ with respect to the norm topology of E. □

The following is a generalization of Lemma 5.1.

Lemma 5.2. *Let $G : [0,1] \times [0,1] \to \mathbf{R}$ be a mapping with the following properties*

(i) for each $t \in [0,1]$, $G(t,.)$ is Lebesgue-measurable on $[0,1]$,
(ii) for each $s \in [0,1]$, $G(.,s)$ is continuous on $[0,1]$,
(iii) there is a constant $M > 0$ such that $|G(t,s)| \le M$ for all $(t,s) \in [0,1] \times [0,1]$.

Let $\Gamma : [0,1] \to E$ be a convex compact valued measurable and Pettis-integrable mapping. Then the set

$$\{u_f : u_f(t) = \int_0^1 G(t,s)f(s)ds : t \in [0,1],\ f \in S_\Gamma^{Pe}\}$$

is equicontinuous in $C_E([0,1])$.

Proof. By Lemma 5.1 it is clear that

$$\{u_f : u_f(t) = \int_0^1 G(t,s)f(s)ds : t \in [0,1],\ f \in S_\Gamma^{Pe}\} \subset C_E([0,1]).$$

Let us check the equicontinuity property. Indeed, let $t, t_k \in [\tau, 1]$ such that $t_k \to t$, we have the estimation

$$||u_f(t) - u_f(t_k)|| \le \sup_{x^* \in \overline{B}_{E^*}} \int_0^1 |G(t_k,s) - G(t,s)||\delta^*(x^*, \Gamma(s))|ds.$$

As the sequence $(|G(t_k,.) - G(t,.)|)$ is bounded in $L^\infty_{\mathbf{R}}([0,1])$ and the set $\{|\delta^*(x^*, \Gamma(.))| : ||x^*|| \le 1\}$ is uniformly integrable in $L^1_{\mathbf{R}}([0,1])$, by invoking again Grothendieck lemma [24] as in the proof of Lemma 5.1, the second term goes to 0 when $t_k \to t$ showing that $\{u_f : f \in S_\Gamma^{Pe}\}$ is equicontinuous in $C_E([0,1])$. □

The following lemma is crucial in the statement of the (SODE) with Pettis-integrable second member and m-point boundary condition. Here we suppose that the hypotheses and notations of Lemma 2.1 hold.

Lemma 5.3. *Let $x \in E$, let G_τ be the Green function, $e_{\tau,x}$ and $\dot{e}_{\tau,x}$ in Lemma 2.1*

$$\begin{cases} e_{\tau,x}(t) = x + A_\tau(1 - \sum_{i=1}^{m-2} \alpha_i)(1 - \exp(-\gamma(t-\tau)))x,\ \forall t \in [\tau, 1] \\ \dot{e}_{\tau,x}(t) = \gamma A_\tau \left(1 - \sum_{i=1}^{m-2} \alpha_i\right) \exp(-\gamma(t-\tau))x,\ \forall t \in [\tau, 1] \\ A_\tau = \left(\sum_{i=1}^{m-2} \alpha_i - 1 + \exp(-\gamma(1-\tau)) - \sum_{i=1}^{m-2} \alpha_i \exp(-\gamma(\eta_i - \tau))\right)^{-1} \end{cases}$$

and let f be a Pettis-integrable function. Let us consider the mapping

$$u_{\tau,x,f}(t) = e_{\tau,x}(t) + \int_\tau^1 G_\tau(t,s)f(s)ds, \quad \tau \in [0,\eta_1[, \quad t \in [0,1].$$

Then the following assertions hold

(1) $u_{\tau,x,f}$ is continuous i.e. $u_{\tau,x,f} \in C_E([0,1])$,
(2) $u_{\tau,x,f}(\tau) = x, \quad u_{\tau,x,f}(1) = \sum_{i=1}^{m-2} \alpha_i u_{\tau,x,f}(\eta_i)$,
(3) The function $u_{\tau,x,f}$ is scalarly derivable, that is, for every $x^* \in E^*$, the scalar function $\langle x^*, u_{\tau,x,f}\rangle$ is derivable and its weak derivative $\dot{u}_{\tau,x,f}$ satisfies

$$\dot{u}_{\tau,x,f}(t) = \dot{e}_{\tau,x}(t) + \int_\tau^1 \frac{\partial G_\tau}{\partial t}(t,s)f(s)ds, \quad \tau \in [0,\eta_1[, \quad t \in [\tau,1].$$

(4) The function $\dot{u}_{\tau,x,f}$ is continuous and scalarly derivable, that is, for every $x^* \in E^*$, the scalar function $\langle x^*, \dot{u}_{\tau,x,f}\rangle$ is derivable and its weak derivative $\ddot{u}_{\tau,x,f}$ satisfies

$$\ddot{u}_{\tau,x,f}(t) + \gamma u_{\tau,x,f}(t) = f(t) \quad a.e. \quad t \in [\tau,1].$$

Proof. (1) Since $e_{\tau,x} \in C_E([0,1])$ and G_τ is a Carathéodory and bounded function, $u_{\tau,x,f}$ is continuous on $[\tau,1]$ with respect to the norm topology of E in view of Lemma 5.1.
(2) follows from Lemma 2.1(iv).
(3)–(4) Similarly, using the property of $\frac{\partial G_\tau}{\partial t}$ in Lemma 2.1 we infer that $t \mapsto \int_\tau^1 \frac{\partial G_\tau}{\partial t}(t,s)f(s)ds$ is continuous on $[\tau,1]$ with respect to the norm topology of E in view of Lemma 5.1 and so is the mapping $t \mapsto \dot{e}_{\tau,x}(t) + \int_\tau^1 \frac{\partial G_\tau}{\partial t}(t,s)f(s)ds$. Now (3)–(4) follow from the computation used in (iv)–(v) in Lemma 2.1. □

By $W_{P,E}^{2,1}([\tau,1])$ we denote the space of all continuous functions in $C_E([\tau,1])$ such that their first weak derivatives are continuous and their second weak derivatives are Pettis-integrable on $[\tau,1]$. By Lemma 5.3, given a Pettis-integrable function $f : [\tau,1] \to E$ (shortly $f \in P_E^1([\tau,1])$) the (SODE)

$$\begin{cases} \ddot{u}_{\tau,x,f}(t) + \gamma \dot{u}_{\tau,x,f}(t) = f(t), t \in [\tau,1], \ \tau \in [0,\eta_1[\\ u_{\tau,x,f}(\tau) = x, \quad u_{\tau,x,f}(1) = \sum_{i=1}^{m-2} \alpha_i u_{\tau,x,f}(\eta_i) \end{cases}$$

admits a unique $W^{2,1}_{P,E}([\tau, 1])$-solution with integral representation formulas

$$u_{\tau,x,f}(t) = e_{\tau,x}(t) + \int_\tau^1 G_\tau(t,s)f(s)ds, \quad \tau \in [0, \eta_1[, \quad t \in [\tau, 1],$$

$$\dot{u}_{\tau,x,f}(t) = \dot{e}_{\tau,x}(t) + \int_\tau^1 \frac{\partial G_\tau}{\partial t}(t,s)f(s)ds, \quad \tau \in [0, \eta_1[, \quad t \in [\tau, 1].$$

The following result provides the compactness of solutions for a class of (SODE) with m ($m > 3$) point boundary condition and Pettis-integrable controls.

Theorem 5.1. *Let E be a separable Banach space and let $\Gamma : [0,1] \to ck(E)$ be a convex compact valued measurable and Pettis-integrable mapping. Let us consider the following*

$$(SODE)_\Gamma \begin{cases} \ddot{u}_{\tau,x,f}(t) + \gamma \dot{u}_{\tau,x,f}(t) = f(t), t \in [\tau, 1], \ \tau \in [0, \eta_1[, \ f \in S^{Pe}_\Gamma \\ u_{\tau,x,f}(\tau) = x, \quad u_{\tau,x,f}(1) = \sum_{i=1}^{m-2} \alpha_i u_{\tau,x,f}(\eta_i). \end{cases}$$

Then the set $\{u_{\tau,x,f} : f \in S^{Pe}_\Gamma\}$ of $W^{2,1}_{P,E}([\tau, 1])$-solutions to $(SODE)_\Gamma$ is compact in $C_E([\tau, 1])$.

Proof. Let (u_{τ,x,f_n}) be a sequence of $W^{2,1}_{P,E}([\tau,1])$-solutions to $(SODE)_\Gamma$. As S^{Pe}_Γ is sequentially $\sigma(P^1_E, L^\infty \otimes E^*)$-compact, by extracting a subsequence we may assume that (f_n) converges with respect to the $\sigma(P^1_E, L^\infty \otimes E^*)$ topology to $f_\infty \in S^{Pe}_\Gamma$. Using Lemma 5.3, we have, for each $n \in \mathbf{N}$,

$$u_{\tau,x,f_n}(t) = e_{\tau,x}(t) + \int_\tau^1 G_\tau(t,s)f_n(s)ds, \ t \in [\tau, 1] \tag{5.1.1}$$

$$\dot{u}_{\tau,x,f_n}(t) = \dot{e}_{\tau,x}(t) + \int_\tau^1 \frac{\partial G_\tau}{\partial t}(t,s)f_n(s)ds, \ t \in [\tau, 1] \tag{5.1.2}$$

$$\ddot{u}_{\tau,x,f_n}(t) + \gamma \dot{u}_{\tau,x,f_n}(t) = f_n(t) \in \Gamma(t), a.e.\ t \in [\tau, 1]. \tag{5.1.3}$$

From the property the Green function G_τ in Lemma 2.1, (5.1.1) and Lemma 5.2, we infer that $\{u_{\tau,x,f_n} : n \in \mathbf{N}\}$ is equicontinuous in $C_E([0, 1])$. Further, for each $t \in [\tau, 1]$, $\{u_{\tau,x,f_n}(t) : n \in \mathbf{N}\}$ is relatively compact because it is included in the norm compact set $e_{\tau,x}(t) + \int_0^1 G_\tau(t,s)\Gamma(s)ds$ (see e.g. [12, 14]). So by Ascoli's theorem, $\{u_{\tau,x,f_n} : n \in \mathbf{N}\}$ is relatively compact in $C_E([\tau, 1])$. Similarly using the properties of $\frac{\partial G_\tau}{\partial t}$ in Lemma 2.1, (5.1.2) and Lemma 5.2, we deduce that $\{\dot{u}_{\tau,x,f_n} : n \in \mathbf{N}\}$ is equicontinuous in $C_E([\tau, 1])$. In addition, the set $\{\dot{u}_{\tau,x,f_n}(t) : n \in \mathbf{N}\}$ is included in the compact set $\dot{e}_{\tau,x}(t) + \int_0^1 \frac{\partial G_\tau}{\partial t}(t,s)\,\Gamma(s)\,ds$. So $\{\dot{u}_{\tau,x,f_n} : n \in \mathbf{N}\}$ is relatively compact in $C_E([\tau, 1])$ using the Ascoli's theorem. From the above facts, we deduce that there exists a subsequence of $(u_{\tau,x,f_n})_{n\in\mathbf{N}}$ still denoted by $(u_{\tau,x,f_n})_{n\in\mathbf{N}}$ which converges uniformly to $u^\infty \in C_E([\tau, 1])$ with $u^\infty(0) = x$ and $u^\infty(1) = \sum_{i=1}^{m-2} \alpha_i u^\infty(\eta_i)$. Similarly, we may assume that $(\dot{u}_{\tau,x,f_n})$ converges uniformly to $v^\infty \in C_E([\tau, 1])$. Furthermore, by the above facts, it is easy to see that $(\ddot{u}_{\tau,x,f_n})$ converges $\sigma(P_E^1, L^\infty \otimes E^*)$ to a Pettis integrable function $w^\infty \in P_E^1([\tau, 1])$. For every $t \in [\tau, 1]$, using the representation formula (5.1.1), we have

$$
\begin{aligned}
u^\infty(t) &\\
&= \lim_{n\to\infty} u_{\tau,x,f_n}(t) = e_{\tau,x}(t) + \lim_{n\to\infty} \int_\tau^1 G_\tau(t,s)(\ddot{u}_{\tau,x,f_n}(s) + \gamma \dot{u}_{\tau,x,f_n}(s))ds \\
&= e_{\tau,x}(t) + \lim_{n\to\infty} \int_0^1 G_\tau(t,s)\ddot{u}_{\tau,x,f_n}(s)ds + \gamma \lim_{n\to\infty} \int_\tau^1 G_\tau(t,s)\dot{u}_{\tau,x,f_n}(s)ds \\
&= e_{\tau,x}(t) + \int_0^1 G_\tau(t,s)w^\infty(s)ds + \gamma \int_0^1 G_0(t,s)v^\infty(s)ds \\
&= e_{\tau,x}(t) + \int_0^1 G_\tau(t,s)(w^\infty(s) + \gamma v^\infty(s))ds. \qquad (5.1.4)
\end{aligned}
$$

From (5.1.4) and Lemma 5.3, we deduce that u^∞ is scalarly derivable and its weak derivative $\dot{u}^\infty$ is given by

$$
\dot{u}^\infty(t) = \dot{e}_{\tau,x}(t) + \int_\tau^1 \frac{\partial G_\tau}{\partial t}(t,s)(w^\infty(s) + \gamma v^\infty(s))ds, \forall t \in [\tau, 1]. \quad (5.1.5)
$$

Now using the integral representation formula (5.1.2) we have, for every $t \in [\tau, 1]$,

$$
\begin{aligned}
v^\infty(t) &= \lim_{n\to\infty} \dot{u}_{\tau,x,f_n}(t) \\
&= \dot{e}_{\tau,x}(t) + \lim_{n\to\infty} \int_\tau^1 \frac{\partial G_\tau}{\partial t}(t,s)(\ddot{u}_{\tau,x,f_n}(s) + \gamma \dot{u}_{\tau,x,f_n}(s))ds \\
&= \dot{e}_{\tau,x}(t) + \lim_{n\to\infty} \int_\tau^1 \frac{\partial G_\tau}{\partial t}(t,s)\ddot{u}_{f_n}(s)ds \\
&\quad + \gamma \lim_{n\to\infty} \int_\tau^1 \frac{\partial G_\tau}{\partial t}(t,s)\dot{u}_{f_n}(s)ds
\end{aligned}
$$

$$= \dot{e}_{\tau,x}(t) + \int_\tau^1 \frac{\partial G_\tau}{\partial t}(t,s)w^\infty(s)ds + \gamma \int_\tau^1 \frac{\partial G_\tau}{\partial t}(t,s)v^\infty(s)ds$$

$$= \dot{e}_{\tau,x}(t) + \int_\tau^1 \frac{\partial G_\tau}{\partial t}(t,s)(w^\infty(s) + \gamma v^\infty(s))ds \tag{5.1.6}$$

so that by (5.1.5) and (5.1.6) we get $v^\infty = \dot{u}^\infty$. Now using (5.1.4) and invoking Lemma 5.3(4) we get

$$\ddot{u}^\infty(t) + \gamma\dot{u}^\infty(t) = w^\infty(t) + \gamma v^\infty(t) = w^\infty(t) + \gamma\dot{u}^\infty(t) \quad a.e. \quad t \in [\tau, 1].$$

Thus we get $\ddot{u}^\infty(t) = w^\infty(t)\, a.e.\, t \in [\tau, 1]$ so that

$$\begin{cases} u^\infty(t) = e_{\tau,x}(t) + \displaystyle\int_\tau^1 G_\tau(t,s)(\ddot{u}^\infty(s) + \gamma\dot{u}^\infty(s))ds, & t \in [\tau, 1] \\ u^\infty(\tau) = x, \quad u^\infty(1) = \displaystyle\sum_{i=1}^{m-2} \alpha_i u^\infty(\eta_i). \end{cases} \tag{5.1.7}$$

Step 2. Main fact: u^∞ coincides with the $W^{2,1}_{P,E}([\tau, 1])$-solution u_{f_∞} associated with $f_\infty \in S^{Pe}_\Gamma$ to

$$\begin{cases} \ddot{u}_{f_\infty}(t) + \gamma\dot{u}_{f_\infty}(t) = f_\infty(t),\ t \in \tau, 1] \\ u_{f_\infty}(\tau) = x, \quad u_{f_\infty}(1) = \displaystyle\sum_{i=1}^{m-2} \alpha_i u_{f_\infty}(\eta_i). \end{cases} \tag{5.1.8}$$

Remember that

$$\begin{cases} \ddot{u}_{\tau,x,f_n}(t) + \gamma\dot{u}_{f_n}(t) = f_n(t), \\ u_{\tau,x,f_n}(\tau) = x, \quad u_{\tau,x,f_n}(1) = \displaystyle\sum_{i=1}^{m-2} \alpha_i u_{\tau,x,f_n}(\eta_i) \end{cases}$$

and by the above fact, $(\ddot{u}_{\tau,x,f_n} + \gamma\dot{u}_{\tau,x,f_n})$ $\sigma(P^1_E, L^\infty \otimes E^*)$-converges in $P^1_E([\tau, 1])$ to $\ddot{u}^\infty + \gamma\dot{u}^\infty$. Let $v = h \otimes x^* \in L^\infty([\tau, 1]) \otimes E^*$. Multiply scalarly the equation

$$\ddot{u}_{\tau,x,f_n}(t) + \gamma\dot{u}_{\tau,x,f_n}(t) = f_n(t)$$

by $v(t)$ and integrating on $[\tau, 1]$ yields

$$\int_\tau^1 \langle h(t) \otimes x^*, \ddot{u}_{f_n}(t) + \gamma\dot{u}_{f_n}(t)\rangle dt = \int_\tau^1 \langle h(t) \otimes x^*, f_n(t)\rangle dt. \tag{5.1.9}$$

It is clear that

$$\lim_{n\to\infty}\int_\tau^1 \langle h(t)\otimes x^*, \ddot{u}_{f_n}(t)+\gamma\dot{u}_{f_n}(t)\rangle dt = \int_\tau^1 \langle h(t)\otimes x^*, \ddot{u}^\infty(t)+\gamma\dot{u}^\infty(t)\rangle dt$$

$$= \lim_{n\to\infty}\int_\tau^1 \langle h(t)\otimes x^*, f_n(t)\rangle dt = \int_\tau^1 \langle h(t)\otimes x^*, f_\infty(t)\rangle dt$$

so that by invoking the separability of E

$$\ddot{u}^\infty(t) + \gamma\dot{u}^\infty(t) = f_\infty(t) \quad a.e. \quad t \in [\tau, 1]. \tag{5.1.10}$$

Using (5.1.7), (5.1.8), and (5.1.10) and uniqueness of solution we obtain $u^\infty = u_{f_\infty}$. □

Remark. In the context of Theory of Control, we have stated in the proof of Theorem 5.1, the dependence of the trajectory solution with respect to the Pettis controls. Namely, with the notations of Theorem 5.1, if u_{τ,x,f_n} is the $W^{2,1}_{P,E}([\tau, 1])$-solution of

$$\begin{cases} \ddot{u}_{\tau,x,f_n}(t) + \gamma\dot{u}_{f_n}(t) = f_n(t), \quad t \in [\tau, 1] \\ u_{\tau,x,f_n}(\tau) = x, \quad u_{\tau,x,f_n}(1) = \sum_{i=1}^{m-2} \alpha_i u_{\tau,x,f_n}(\eta_i) \end{cases}$$

and if (f_n) $\sigma(P^1_E, L^\infty \otimes E^*)$-converges to $f_\infty \in S^{Pe}_\Gamma$, then (u_{τ,x,f_n}) converges uniformly to u_{τ,x,f_∞}, $(\dot{u}_{\tau,x,f_n})$ converges uniformly to $\dot{u}_{\tau,x,f_\infty}$ and $(\ddot{u}_{\tau,x,f_n})$ $\sigma(P^1_E, L^\infty \otimes E^*)$-converges to $\ddot{u}_{\tau,x,f_\infty}$ where u_{τ,x,f_∞} is the $W^{2,1}_{P,E}([\tau, 1])$-solution of

$$\begin{cases} \ddot{u}_{\tau,x,f_\infty}(t) + \gamma\dot{u}_{\tau,x,f_\infty}(t) = f_\infty(t), \quad t \in [\tau, 1] \\ u_{\tau,x,f_\infty}(\tau) = x, \quad u_{\tau,x,f_\infty}(1) = \sum_{i=1}^{m-2} \alpha_i u_{\tau,x,f_\infty}(\eta_i). \end{cases}$$

The above remark is of importance since it allows to prove further results. Here is an application to the existence of $W^{2,1}_{P,E}([\tau, 1])$-solution of a (SODE) with m-point boundary condition.

Theorem 5.2. *Let $F : [0, 1] \times (E \times E) \to E$ be a Carathéodory mapping satisfying*

$$F(t, x, y) \in \Gamma(t)$$

for all $(t, x, y) \in [0, 1] \times E \times E$ *where* $\Gamma : [0, 1] \rightrightarrows E$ *is a convex compact valued Pettis-integrable mapping. Then the (SODE)*

$$\begin{cases} \ddot{u}(t) + \gamma \dot{u}(t) = F(t, u(t), \dot{u}(t)), \quad t \in [\tau, 1] \\ u(\tau) = x, \quad u(1) = \sum_{i=1}^{m-2} \alpha_i u(\eta_i) \end{cases}$$

has a $W^{2,1}_{P,E}([\tau, 1])$*-solution.*

Proof. Let us set

$$\mathcal{X} := \{u_{\tau,x,f} : [\tau, 1] \to E : u_{\tau,x,f}(t) = e_{\tau,x}(t) + \int_\tau^1 G_\tau(t, s) f(s) ds, \ t \in [\tau, 1], \ f \in S^{Pe}_\Gamma\}.$$

Then Theorem 5.1 shows that $\mathcal{X}$ is compact and convex in $C_E([\tau, 1])$. For each $u \in \mathcal{X}$, let us set

$$\Phi(u) := \{w \in \mathcal{X} : \ddot{w}(t) + \gamma \dot{w}(t) = F(t, u(t), \dot{u}(t)), \ t \in [\tau, 1]\}.$$

In view of Lemma 5.3, $\Phi(u)$ is non empty. Let us prove that the mapping $\Phi : \mathcal{X} \to \mathcal{X}$ is continuous. Let $(u_n, v_n) \in \text{Graph}\,\Phi$ such that $u_n \to u$ and $v_n \to v$ in $\mathcal{X}$. We need to check that $v = \Phi(u)$. Taking account of the particular structure of $\mathcal{X}$ and the remark of Theorem 5.1, we have that $\dot{u}_n \to \dot{u}$ uniformly and $\ddot{u}_n$ $\sigma(P^1_E, L^\infty \otimes E^*)$-converges to $\ddot{u}$ and that $\dot{v}_n \to \dot{v}$ uniformly and $\ddot{v}_n$ $\sigma(P^1_E, L^\infty \otimes E^*)$-converges to $\ddot{v}$. Multiply scalarly the equality

$$\ddot{v}_n(t) + \gamma \dot{v}_n(t) = F(t, u_n(t), \dot{u}_n(t)), \quad t \in [\tau, 1]$$

by $h(t) \otimes x^*$ where $h \in L^\infty_{\mathbf{R}^+}([\tau, 1])$ and $x^* \in \overline{B}_{E^*}$ and integrating on $[\tau, 1]$ gives

$$\int_\tau^1 \langle h(t) \otimes x^*, \ddot{v}_n(t) + \gamma \dot{v}_n(t) \rangle dt = \int_\tau^1 \langle h(t) \otimes x^*, F(t, u_n(t), \dot{u}_n(t)) \rangle dt. \tag{5.2.1}$$

By passing to the limit when $n \to \infty$ in (5.2.1) we get

$$\lim_{n\to\infty} \int_\tau^1 \langle h(t) \otimes x^*, \ddot{v}_n(t) + \gamma \dot{v}_n(t) \rangle dt = \int_\tau^1 \langle h(t) \otimes x^*, \ddot{v}(t) + \gamma \dot{v}(t) \rangle dt$$
$$= \lim_{n\to\infty} \int_\tau^1 \langle h(t) \otimes x^*, F(t, u_n(t), \dot{u}_n(t)) \rangle dt = \int_\tau^1 \langle h(t) \otimes x^*, F(t, u(t), \dot{u}(t)) \rangle dt \tag{5.2.2}$$

by Lebesgue dominated convergence theorem, because

$$|\langle h(t) \otimes x^*, F(t, x, y)\rangle| \leq h(t)|\delta^*(x^*, \Gamma(t))|$$

for all $(t, x, y) \in [0, 1] \times E \times E$. By (5.2.2) we deduce that

$$\int_\tau^1 \langle h(t) \otimes x^*, \ddot{v}(t) + \gamma \dot{v}(t)\rangle dt = \int_\tau^1 \langle h(t) \otimes x^*, F(t, u(t), \dot{u}(t))\rangle dt.$$

Whence we get

$$\langle x^*, \ddot{v}(t) + \gamma \dot{v}(t)\rangle = \langle x^*, F(t, u(t), \dot{u}(t))\rangle, \quad a.e. \tag{5.2.4}$$

for every $x^* \in \overline{B}_{E^*}$. By taking a dense sequence $(e_k^*)_{k \in \mathbf{N}}$ in $\overline{B}_{E^*}$ for the Mackey topology we get

$$\langle e_k^*, \ddot{v}(t) + \gamma \dot{v}(t)\rangle = \langle e_k^*, F(t, u(t), \dot{u}(t))\rangle, \quad a.e. \tag{5.2.5}$$

for all $k \in \mathbf{N}$. Finally we get

$$\ddot{v}(t) + \gamma \dot{v}(t) = F(t, u(t), \dot{u}(t)), \quad a.e.$$

proving that Graph Φ is compact. By applying the Kakutani–Ky Fan fixed point theorem to Φ, we find $u \in \mathcal{X}$ such that $u = \Phi(u)$ which is a $W^{2,1}_{P,E}([\tau, 1])$-solution of the (SODE) under consideration. □

The compactness in $C_E([\tau, 1])$ of

$$\begin{aligned}\mathcal{X} := \{u_{\tau,x,f} : [\tau, 1] \to E : u_{\tau,x,f}(t) = e_{\tau,x}(t) \\ + \int_\tau^1 G_\tau(t, s) f(s) ds, \ t \in [\tau, 1], \ f \in S_\Gamma^{Pe}\} \\ \mathcal{Y} := \{\dot{u}_{\tau,x,f} : [\tau, 1] \to E : \dot{u}_{\tau,x,f}(t) = \dot{e}_{\tau,x}(t) \\ + \int_\tau^1 \frac{\partial G_\tau}{\partial t}(t, s) f(s) ds, \ t \in [\tau, 1], \ f \in S_\Gamma^{Pe}\}\end{aligned}$$

are of importance and rely on some delicate arguments in the pioneering work of [1, 2] involving the Pettis uniformly integrable condition, Grothendieck lemma characterizing the Mackey topology for bounded sets in $L^\infty_{\mathbf{R}}$ [24] and other compactness results. Second order differential inclusions with three point boundary condition in case where the second member is a Pettis-integrable convex compact valued multifunction is initiated in [2]. At this point a second order differential inclusion with upper semicontinuous convex compact valued multifunction and three point boundary condition of the form

$$\begin{cases} \ddot{u}(t) \in F(t, u(t), \dot{u}(t)) \ a.e \ t \in [0, 1], \\ u(0) = 0; \ \ u(\theta) = u(1). \end{cases}$$

is available in [2, 27]. Taking account into the above facts, one may state the validity of Theorem 5.2 when F is a convex compact valued upper semicontinuous mapping. Since we don't focus on differential inclusion in the paper, we only mention a closure type lemma which may have an independent interest and solves this problem.

Theorem 5.3. *Let $F : [0, 1] \times E \times E \to E$ be a convex compact valued upper semicontinuous mapping satisfying*

$$F(t, x, y) \subset \Gamma(t)$$

for all $(t, x, y) \in [0, 1] \times E \times E$ where $\Gamma : [0, 1] \rightrightarrows E$ is a convex compact valued Pettis-integrable mapping. Let $(u_n, v_n) \in \mathcal{X} \times \mathcal{X}$ such that $u_n \to u$ and $v_n \to v$ in $\mathcal{X}$ and that

$$\ddot{v}_n(t) + \gamma v_n(t) \in F(t, u_n(t), \dot{u}_n(t))$$

for all $n \in \mathbf{N}$ and for all $t \in [\tau, 1]$. Then we have $\ddot{v}(t) + \gamma v(t) \in F(t, u(t), \dot{u}(t))$ a.e.

Proof. Let $h \otimes x^*$ where $h \in L^{\infty}_{\mathbf{R}^+}([\tau, 1])$ and $x^* \in \overline{B}_{E^*}$. From

$$\ddot{v}_n(t) + \gamma v_n(t) \in F(t, u_n(t), \dot{u}_n(t))$$

we have

$$\langle h(t) \otimes x^*, \ddot{v}_n(t) + \gamma v_n(t) \rangle \leq \delta^*(h(t) \otimes x^*, F(t, u_n(t), \dot{u}_n(t))).$$

Integrating on $[\tau, 1]$ this inequality yields

$$\int_{\tau}^{1} \langle h(t) \otimes x^*, \ddot{v}_n(t) + \gamma v_n(t) \rangle dt \leq \int_{\tau}^{1} \delta^*(h(t) \otimes x^*, \times F(t, u_n(t), \dot{u}_n(t))) dt. \quad (5.3.1)$$

Repeating the arguments of the proof of Theorem 5.2, we have that $\dot{u}_n \to \dot{u}$ uniformly and $\ddot{u}_n$ $\sigma(P^1_E, L^{\infty} \otimes E^*)$-converges to $\ddot{u}$ and that $\dot{v}_n \to \dot{v}$ uniformly and $\ddot{v}_n$ $\sigma(P^1_E, L^\infty \otimes E^*)$-converges to $\ddot{v}$. Then by passing to the limit when $n \to \infty$ in (5.3.1) we get

$$\begin{aligned}
\lim_{n\to\infty} \int_\tau^1 &\langle h(t) \otimes x^*, \ddot{v}_n(t) + \gamma \dot{v}_n(t)\rangle dt \\
&= \int_\tau^1 \langle h(t) \otimes x^*, \ddot{v}(t) + \gamma \dot{v}(t)\rangle dt \\
&\leq \limsup_{n\to\infty} \int_\tau^1 h(t)\delta^*(x^*, F(t, u_n(t), \dot{u}_n(t)))dt \\
&\leq \int_\tau^1 h(t) \limsup_{n\to\infty} \delta^*(x^*, F(t, u_n(t), \dot{u}_n(t)))dt \\
&\leq \int_\tau^1 h(t)\delta^*(x^*, F(t, u(t), \dot{u}(t)))dt \qquad (5.3.2)
\end{aligned}$$

because

$$|\delta^*(h(t) \otimes x^*, F(t, x, y))| \leq |\delta^*(h(t) \otimes x^*, \Gamma(t))| = h(t)|\delta^*(x^*, \Gamma(t))|$$

for all $(t, x, y) \in [0, 1] \times E \times E$ and the mapping F is upper semicontinuous. By (5.3.2) we deduce that

$$\int_\tau^1 h(t)\langle x^*, \ddot{v}(t) + \gamma \dot{v}(t)\rangle dt \leq \int_\tau^1 h(t)\delta^*(x^*, F(t, u(t), \dot{u}(t)))dt.$$

Whence we get

$$\langle x^*, \ddot{v}(t) + \gamma \dot{v}(t)\rangle \leq \delta^*(x^*, F(t, u(t), \dot{u}(t))) \quad a.e.$$

for every $x^* \in \overline{B}_{E^*}$. By taking a dense sequence $(e^*_k)_{k\in\mathbf{N}}$ in $\overline{B}_{E^*}$ for the Mackey topology we get

$$\langle e^*_k, \ddot{v}(t) + \gamma \dot{v}(t)\rangle \leq \delta^*(e^*_k, F(t, u(t), \dot{u}(t))) \quad a.e.$$

for all $k \in \mathbf{N}$ so that

$$\ddot{v}(t) + \gamma \dot{v}(t) \in F(t, u(t), \dot{u}(t)) \quad a.e.$$

□

6. Open Problems: Differential Game Governed by (SODE), (ODE) and Sweeping Process with Strategies

To finish the paper we discuss some viscosity problems in a differential game governed by a class of (ODE) with strategy in the line of Elliot [20], Elliot–Kalton [21] and Evans–Souganides [22]. For simplicity we assume that E is a separable Hilbert space. Let us consider two compact subsets Y and Z in E. Set

$$Y(\tau) = \{y : [\tau, 1] \to Y |\, y \text{ measurable}\}$$
$$Z(\tau) = \{z : [\tau, 1] \to Z |\, z \text{ measurable}\}$$

Denote by $\Gamma(\tau)$ the set of all strategies $\alpha : Z(\tau) \to Y(\tau)$ and $\Delta(\tau)$ the set of all strategies $\beta : Y(\tau) \to Z(\tau)$. Let us given a Carathéodory integrable mapping $F : [0,1] \times (Y \times Z) \to E$ such that $F(t, y, z) \subset K(t)$ for all $(t, y, z) \in [0, 1] \times Y \times Z$ where $K : [0, 1] \Rightarrow E$ is a convex compact valued integrably bounded mapping, a bounded continuous integrand $J : [0, 1] \times E \times Y \times Z \to \mathbf{R}$ and let us define the upper–lower value function

$$U_J(\tau, x) = \sup_{\alpha \in \Gamma(\tau)} \inf_{z \in Z(\tau)} \{\int_\tau^1 J(t, u_{\tau,x,\alpha(z),z}(t), \alpha(z)(t), z(t))dt\}, \tau \in [0, \eta_1]$$

$$V_J(\tau, x) = \inf_{\beta \in \Delta(\tau)} \sup_{y \in Y(\tau)} \{\int_\tau^1 J(t, u_{\tau,x,y,\beta(y)}(t), y(t), \beta(y)(t))dt\}, \tau \in [0, \eta_1]$$

where $u_{\tau,x,\alpha(z),z}$ is the trajectory $W_E^{2,1}([\tau, 1])$-solution of second order differential game

$$\begin{cases} \ddot{u}_{\tau,x,\alpha(z),z}(t) + \gamma \dot{u}_{\tau,x,\alpha(z),z}(t) = F(t, \alpha(z)(t), z(t)), t \in [\tau, 1], \ \tau \in [0, \eta_1] \\ u_{\tau,x,\alpha(z),z}(\tau) = x, \\ u_{\tau,x,\alpha(z),z}(1) = \sum_{i=1}^{m-2} \alpha_i u_{\tau,x,\alpha(z),z}(\eta_i), \end{cases} \tag{6.1.1}$$

with the integral representation formulas

$$u_{\tau,x,\alpha(z),z}(t)(t) = e_{\tau,x}(t) + \int_\tau^1 G_\tau(t, s)F(s, \alpha(z)(s), z(s))ds, \ \ t \in [\tau, 1]$$

$$\dot{u}_{\tau,x,\alpha(z),z}(t) = \dot{e}_{\tau,x}(t) + \int_\tau^1 \frac{\partial G_\tau}{\partial t}(t, s)F(s, \alpha(z)(s), z(s))ds, \ \ t \in [\tau, 1]$$

and similarly $u_{\tau,x,y,\beta(y)}$ is the trajectory $W_E^{2,1}([\tau, 1])$-solution of second order differential game

$$\begin{cases} \ddot{u}_{\tau,x,y,\beta(y)}(t) + \gamma \ddot{u}_{\tau,x,y,\beta(y)}(t) = F(t, y(t), \beta(y)(t)), t \in [\tau, 1], \tau \in [0, \eta_1] \\ u_{\tau,x,y,\beta(y)}(\tau) = x, \\ u_{\tau,x,y,\beta(y)}(1) = \sum_{i=1}^{m-2} \alpha_i u_{\tau,x,y,\beta(y)}(\eta_i). \end{cases} \tag{6.1.2}$$

We aim to generalize the viscosity problem in Theorem 4.2 to the case of strategies in the following

Proposition 6.1. *Let* $J : [0,1] \times E \times Y \times Z \to \mathbf{R}$ *be a bounded continuous integrand,* $\tau, \sigma \in [0,1]$ *such that* $\tau \in [0, \eta_1[$ *and* $\tau + \sigma < 1$ *and let us consider the upper value function*

$$U_J(\tau, x) = \sup_{\alpha \in \Gamma(\tau)} \inf_{z \in Z(\tau)} \left\{ \int_\tau^1 J(t, u_{\tau,x,\alpha(z),z}(t), \alpha(z)(t), z(t))dt \right\},$$
$$\tau \in [0, \eta_1], x \in E.$$

where $u_{\tau,x,\alpha(z),z}$ *is the trajectory* $W_E^{2,1}([\tau, 1])$*-solution of second order differential game*

$$\begin{cases} \ddot{u}_{\tau,x,\alpha(z),z}(t) + \gamma \dot{u}_{\tau,x,\alpha(z),z}(t) = F(t, \alpha(z)(t), z(t)), \ t \in [\tau, 1], \ \tau \in [0, \eta_1] \\ u_{\tau,x,\alpha(z),z}(\tau) = x, \\ u_{\tau,x,\alpha(z),z}(1) = \sum_{i=1}^{m-2} \alpha_i u_{\tau,x,\alpha(z),z}(\eta_i). \end{cases} \tag{6.1.1}$$

Then U_J *satisfies a sub-viscosity property: For any* $\varphi \in C^1([0,1] \times E)$ *such that* $U_J - \varphi$ *reaches a local maximum at* $(t_0, x_0) \in [0, \eta_1[\times E$*, then*

$$\frac{\partial \varphi}{\partial t}(t_0, x_0) + \min_{z \in Z} \max_{y \in Y} \{J(t_0, x_0, y, z)\} + \delta^*(\nabla\varphi(t_0, x_0), \dot{e}_{t_0,x_0}(t_0)$$
$$+ \int_{t_0}^1 \frac{\partial G_{t_0}}{\partial t}(t_0, s)K(s)ds) \geq 0$$

provides that U_J *satisfies the DPP*

$$U_J(\tau, x) = \sup_{\alpha \in \Gamma(\tau)} \inf_{z \in Z(\tau)} \left\{ \int_\tau^{\tau+\sigma} J(t, u_{\tau,x,\alpha(z),z}(s), \alpha(z)(s), z(s))ds \right.$$
$$\left. + U_J(\tau + \sigma), u_{\tau,x,\alpha(z),z}(\tau + \sigma) \right\}.$$

Proof. Assume there is a $\varphi \in C^1([0,1] \times E)$ such that $U_J - \varphi$ reaches a local maximum at $(t_0, x_0) \in [0, \eta_1[\times E$ for which

$$\frac{\partial \varphi}{\partial t}(t_0, x_0) + \min_{z \in Z} \max_{y \in Y} \{J(t_0, x_0, y, z)\} + \delta^*(\nabla\varphi(t_0, x_0), \dot{e}_{t_0,x_0}(t_0)$$
$$+ \int_{t_0}^1 \frac{\partial G_{t_0}}{\partial t}(t_0, s)K(s)ds) < 0.$$

Hence there exists some $\eta > 0$ such that

$$\frac{\partial \varphi}{\partial t}(t_0, x_0) + \min_{z\in Z} \max_{y\in Y}\{J(t_0, x_0, y, z)\} + \delta^*(\nabla\varphi(t_0, x_0), \dot{e}_{t_0,x_0}(t_0) + \int_{t_0}^{1} \frac{\partial G_{t_0}}{\partial t}(t_0, s)K(s)ds) \leq -\eta < 0.$$

Set

$$\Lambda(t, x, y, z) = \frac{\partial \varphi}{\partial t}(t, x) + J(t, x, y, z) + \delta^*(\nabla\varphi(t, x), \dot{e}_{t_0,x_0}(t) + \int_{t_0}^{1} \frac{\partial G_{t_0}}{\partial t}(t, s)K(s)ds).$$

Then we have

$$\min_{z\in Z} \max_{y\in Y} \Lambda(t_0, x_0, y, z) \leq -\eta < 0.$$

Hence there exists some $\overline{z} \in Z$ such that

$$\max_{y\in Y} \Lambda(t_0, x_0, y, \overline{z}) \leq -\eta < 0.$$

Since the mapping

$$(t, x) \mapsto \max_{y\in Y} \Lambda(t_0, x_0, y, \overline{z})$$

is continuous there is $\varepsilon > 0$ such that

$$\max_{y\in Y} \Lambda(t, x, y, \overline{z}) < -\frac{\eta}{2}$$

for $0 \leq t - t_0 \leq \varepsilon$ and $||x - x_0|| \leq \varepsilon$. As $\dot{u}_{t_0,x_0,\alpha(z),z}$ is estimated by

$$||\dot{u}_{t_0,x_0,\alpha(z),z}(t)|| \leq ||\dot{e}_{t_0,x_0}(t)|| + \int_{t_0}^{1} |\frac{\partial G_{t_0}}{\partial t}(t, s)||K(s)|ds = c(t)$$

with $c \in C_{\mathbf{R}}([t_0, 1])$ for all $z \in Z(t_0)$ and for all $\alpha \in \Gamma(t_0)$ in view of the above integral representation formula, so we can choose $\sigma > 0$ such that $||u_{t_0,x_0,\alpha(z),z}(t) - u_{t_0,x_0,\alpha(z),z}(t_0)|| \leq \int_{t_0}^{t_0+\sigma} c(t)dt \leq \varepsilon$ for all $t \in [t_0, t_0 + \sigma]$ and for all $z \in Z(t_0)$ and for all $\alpha \in \Delta(t_0)$. Then the constant control $\overline{z}(t) = \overline{z}$, $\forall t \in [t_0, 1]$ belongs to $Z(t_0)$ and $\alpha(\overline{z})$ belongs to $Y(t_0)$ for all $\alpha \in \Gamma(t_0)$ so that by integrating we have

$$\int_{t_0}^{t_0+\sigma} \Lambda(t, u_{t_0,x_0,\alpha(\overline{z}),\overline{z}}(t), \alpha(\overline{z})(t), \overline{z}(t))dt < -\frac{\sigma\eta}{2}$$

for all $\alpha \in \Gamma(t_0)$. Thus

$$\sup_{\alpha\in\Gamma(t_0)} \left\{ \int_{t_0}^{t_0+\sigma} J(t, u_{t_0,x_0,\alpha(\overline{z}),\overline{z}}(t), \alpha(\overline{z})(t), \overline{z}(t))dt \right. \tag{6.1.3}$$

$$+\delta^*(\nabla\varphi(t, u_{t_0,x_0,\alpha(\overline{z}),\overline{z}}(t)), \dot{e}_{t_0,x_0}(t) + \int_{t_0}^{1} \frac{\partial G_{t_0}}{\partial t}(t,s)K(s)ds)$$

$$\left. + \frac{\partial\varphi}{\partial t}(t, u_{t_0,x_0,\alpha(\overline{z}),\overline{z}}(t) \right\} < -\frac{\sigma\eta}{2}.$$

As U_J satisfies the DPP property, we have

$$U_J(t_0, x_0) \leq \sup_{\alpha\in\Gamma(t_0)} \left\{ \int_{t_0}^{t_0+\sigma} J(t, u_{t_0,x_0,\alpha(\overline{z}),\overline{z}}(t), \alpha(\overline{z})(t), \overline{z}(t))dt \right.$$

$$\left. + U_J(t_0+\sigma, u_{t_0,x_0,\alpha(\overline{z}),\overline{z}}(t_0+\sigma)) \right\}.$$

Hence, for every $n \in \mathbf{N}$, there exists $\alpha^n \in \Gamma(t_0)$ such that

$$U_J(t_0, x_0) \leq \int_{t_0}^{t_0+\sigma} J(t, u_{t_0,x_0,\alpha^n(\overline{z}),\overline{z}}(t), \alpha^n(\overline{z})(t), \overline{z}(t))dt$$

$$+ U_J(t_0+\sigma, u_{t_0,x_0,\alpha^n(\overline{z}),\overline{z}}(t_0+\sigma)) + \frac{1}{n}. \tag{6.1.4}$$

But $U_J - \varphi$ has a local maximum at (t_0, x_0), for small enough σ

$$U_J(t_0, x_0) - \varphi(t_0, x_0) \geq U_J(t_0+\sigma, u_{t_0,x_0,\alpha(z),z}(t_0+\sigma))$$
$$-\varphi(t_0+\sigma, u_{t_0,x_0,\alpha(z),z}(t_0+\sigma)) \tag{6.1.5}$$

for any trajectory solution $u_{t_0,x_0,\alpha(z),z}$ associated with control $(\alpha(z), z)$ ($\alpha \in \Gamma(t_0)$, $z \in Z$). From (6.1.4) and (6.1.5) we deduce

$$U_J(t_0+\sigma, u_{t_0,x_0,\alpha^n(\overline{z}),\overline{z}}(t_0+\sigma)) - \varphi(t_0+\sigma, u_{t_0,x_0,\alpha^n(\overline{z}),\overline{z}}(t_0+\sigma))$$

$$\leq \int_{t_0}^{t_0+\sigma} J(t, u_{t_0,x_0,\alpha^n(\overline{z}),\overline{z}}(t), \alpha^n(\overline{z})(t), \overline{z}(t))dt$$

$$+U_J(t_0+\sigma, u_{t_0,x_0,\alpha^n(\overline{z}),\overline{z}}(t_0+\sigma) + \frac{1}{n} - \varphi(t_0, x_0).$$

Thus we have

$$0 \le \int_{t_0}^{t_0+\sigma} J(t, u_{t_0,x_0,\alpha^n(\overline{z}),\overline{z}}(t), \alpha^n(\overline{z})(t), \overline{z}(t))dt$$
$$+ \varphi(t_0+\sigma, u_{t_0,x_0,\alpha^n(\overline{z}),\overline{z}}(t_0+\sigma)) - \varphi(t_0, x_0) + \frac{1}{n}. \tag{6.1.6}$$

But

$$\begin{aligned}\varphi(t_0+\sigma, u_{t_0,x_0,\alpha^n(\overline{z}),\overline{z}}(t_0+\sigma)) - \varphi(t_0, x_0)\\ = \int_{t_0}^{t_0+\sigma} \langle \nabla\varphi(t, u_{t_0,x_0,\alpha^n(\overline{z}),\overline{z}}(t)), \dot{u}_{t_0,x_0,\alpha^n(\overline{z}),\overline{z}}(t)\rangle dt\\ + \int_{t_0}^{t_0+\sigma} \frac{\partial\varphi}{\partial t}(t, u_{t_0,x_0,\alpha^n(\overline{z}),\overline{z}}(t))dt\end{aligned} \tag{6.1.7}$$

and

$$\dot{u}_{t_0,x_0,\alpha^n(\overline{z}),\overline{z}}(t) = \dot{e}_{t_0,x_0}(t) + \int_{t_0}^{1} \frac{\partial G_{t_0}}{\partial t}(t,s)F(s, \alpha^n(\overline{z})(s), \overline{z}(s))ds$$

because $u_{t_0,x_0,\alpha^n(\overline{z}),\overline{z}}$ is the $W^{2,1}([\tau, 1])$-solution to (SODE)

$$\begin{aligned}\ddot{u}_{t_0,x_0,\alpha^n(\overline{z}),\overline{z}}(t) + \gamma\dot{u}_{t_0,x_0,\alpha^n(\overline{z}),\overline{z}}(t) = F(t, \alpha^n(\overline{z})(t), \overline{z}(t)),\\ u_{t_0,x_0,\alpha^n(\overline{z}),\overline{z}}(t_0) = x_0,\\ u_{t_0,x_0,\alpha^n(\overline{z}),\overline{z}}(1) = \sum_{i=1}^{m-2} \alpha_i u_{t_0,x_0,\alpha^n(\overline{z}),\overline{z}}(\eta_i).\end{aligned}$$

From (6.1.6) and (6.1.7) we deduce

$$\begin{aligned}0 \le \int_{t_0}^{t_0+\sigma} J(t, u_{t_0,x_0,\alpha^n(\overline{z}),\overline{z}}(t), \alpha^n(\overline{z})(t), \overline{z}(t))dt\\ + \int_{t_0}^{t_0+\sigma} \delta^*(\nabla\varphi(t, u_{t_0,x_0,\alpha^n(\overline{z})(t),\overline{z}}(t)), \dot{e}_{t_0,x_0}(t) + \int_{t_0}^{1} \frac{\partial G_{t_0}}{\partial t}(t,s)K(s)ds)dt\\ + \int_{t_0}^{t_0+\sigma} \frac{\partial\varphi}{\partial t}(t, u_{t_0,x_0,\alpha^n(\overline{z}),\overline{z}}(t)) + \frac{1}{n}.\end{aligned} \tag{6.1.8}$$

Using (6.1.3) and (6.1.8) it follows that $0 < \frac{\sigma\eta}{2} < \frac{1}{n}$ for every $n \in \mathbf{N}$. Passing to the limit when n goes to ∞ in the preceding inequality yields a contradiction. □

The viscosity property for the lower–upper value function is an open problem in the present context. Proposition 6.1 is a step forward in the problem under consideration. Compare with earlier result in the literature dealing with viscosity problem governed by (ODE) in $\mathbf{R}^n$ involving differential games and strategies, e.g. [4, 20, 22], evolution inclusions, e.g. [7, 8, 13–17] involving Young control measures, and Relaxation and Bolza problems governed by (SODE), e.g. [3, 9–11]. In order to illustrate the comparison, let us come back to a differential game governed by ordinary differential equation (ODE). Let $\mathcal{M}^1_+(Y)$ and $\mathcal{M}^1_+(Z)$ be the set of all probability Radon measures on compact metric space Y and Z, respectively, endowed with the narrow topology so that $\mathcal{M}^1_+(Y)$ and $\mathcal{M}^1_+(Z)$ are compact metrizable. Consider the space of Young measures (alias relaxed controls)

$$\mathcal{Y}(\tau) = \{y : [\tau, 1] \to \mathcal{M}^1_+(Y) \,|\, y \ \text{ measurable}\}$$

$$\mathcal{Z}(\tau) = \{z : [\tau, 1] \to \mathcal{M}^1_+(Z) \,|\, z \ \text{ measurable}\}$$

and as above denote by $\Gamma(\tau)$ the set of all strategies $\alpha : \mathcal{Z}(\tau) \to \mathcal{Y}(\tau)$ and $\Delta(\tau)$ the set of all strategies $\beta : \mathcal{Y}(\tau) \to \mathcal{Z}(\tau)$. Let $J : [0, 1] \times (E \times Y \times Z) \to \mathbf{R}$ be a bounded Carathéodory integrand and let $F : [0, 1] \times (E \times Y \times Z) \to E$ be a Carathéodory mapping satisfying $F(t, x, y, z) \in K(t)$ for all $(t, x, y, z) \in [0, 1] \times E \times Y \times Z$ where $K : [0, 1] \rightrightarrows E$ is a convex compact valued integrably bounded mapping and a Lipschitz type condition $||F(t, x_1, y, z) - F(t, x_2, y, z)|| \leq \lambda ||x_1 - x_2||$ for all (t, x_1, y, z), (t, x_2, y, z) in $[0, 1] \times E \times Y \times Z$. Then one may consider the lower value function

$$V_J(\tau, x) = \inf_{\beta \in \Delta(\tau)} \sup_{\mu \in \mathcal{Y}(\tau)} \left\{ \int_\tau^1 \left[\int_Z \left[\int_Y J(t, u_{\tau,x,\mu,\beta(\mu)}(t), y, z) \mu_t(dy) \right] \times \beta(\mu)_t(dz) \right] dt \right\}$$

where $u_{\tau,x,\mu,\beta(\mu)}$ is the absolutely continuous solution to (ODE)

$$\dot{u}_{\tau,x,\mu,\beta(\mu)}(t) = \int_Z \left[\int_Y F(t, u_{\tau,x,\mu,\beta(\mu)}(t), y, z) \mu_t(dy) \right] \beta(\mu)_t(dz), \, t \in [\tau, 1]$$

$$u_{\tau,x,\mu,\beta(\mu)}(\tau) = x$$

and the upper value function

$$U_J(\tau, x) = \sup_{\alpha \in \Gamma(\tau)} \inf_{\nu \in \mathcal{Z}(\tau)} \left\{ \int_\tau^1 \left[\int_Z \left[\int_Y J(t, u_{\tau,x,\alpha(\nu),\nu}(t), y, z) \alpha(\nu)_t(dy) \right] \nu_t(dz) \right] \right\} dt,$$
$$\tau \in [0, 1], x \in E$$

where $u_{\tau,x,\alpha(\nu),\nu}$ is the absolutely continuous solution to (ODE)

$$\dot{u}_{\tau,x,\alpha(\nu),\nu}(t) = \int_Z \left[\int_Y F(t, u_{\tau,x,\alpha(\nu)(t),\nu}(t), y, z)\alpha(\nu)_t(dy) \right] \nu_t(dz), t \in [\tau, 1]$$

$$u_{\tau,x,\alpha(\nu),\nu}(\tau) = x$$

and state the viscosity properties for these functions. In the sequel, we will make some additional assumptions on J and F, namely, J and F are continuous and the family $(J(.,.,y,z))_{(y,z)\in Y\times Z}$ is equicontinuous and the family $(F(.,.,y,z))_{(y,z)\in Y\times Z}$ is equicontinuous.

Proposition 6.2. *Let $J : [0,1] \times E \times Y \times Z \to \mathbf{R}$ be a bounded continuous integrand, and let us consider the upper value function*

$$U_J(\tau, x) = \sup_{\alpha\in\Gamma(\tau)} \inf_{\nu\in\mathcal{Z}(\tau)} \left\{ \int_\tau^1 \left[\int_Z \left[\int_Y J(t, u_{\tau,x,\alpha(\nu),\nu}(t), y, z)\alpha(\nu)_t(dy) \right] \nu_t(dz) \right] \right\} dt,$$
$$\tau \in [0,1], x \in E.$$

Let us consider the Hamiltonian

$$H^+(t,x,\rho) = \min_{\nu\in\mathcal{M}_+^1(Z)} \max_{\mu\in\mathcal{M}_+^1(Y)} \left\{ \langle \rho, \int_Z \left[\int_Y F(t,x,y,z)d\mu(y) \right] d\nu(z) \right.$$
$$\left. + \int_Z \left[\int_Y J(t,x,y,z)d\mu(y) \right] d\nu(z) \right\}.$$

Then U_J is a viscosity solution to the HJB equation $\frac{\partial U}{\partial t} + H^+(t,x,\nabla U)) = 0$, that is, for any $\varphi \in C^1([0,1]\times E)$ for which $U_J - \varphi$ reaches a local maximum at $(t_0, x_0) \in [0,1] \times E$ we have

$$\frac{\partial \varphi}{\partial t}(t_0, x_0) + H^+(t_0, x_0, \nabla\varphi(t_0, x_0)) \geq 0$$

and for any $\varphi \in C^1([0,1] \times E)$ for which $U_J - \varphi$ reaches a local minimum at $(t_0, x_0) \in [0,1] \times E$, we have

$$\frac{\partial \varphi}{\partial t}(t_0, x_0) + H^+(t_0, x_0, \nabla\varphi(t_0, x_0)) \leq 0$$

provided that U_J satisfies the DPP

$$U_J(\tau, x) = \sup_{\alpha\in\Gamma(\tau)} \inf_{\nu\in\mathcal{Z}(\tau)} \left\{ \int_\tau^{\tau+\sigma} \left[\int_Z \left[\int_Y J(t, u_{\tau,x,\alpha(\nu),\nu}(s), y, z) \right.\right.\right.$$
$$\left.\left. \times \alpha(\nu)_s(dy) \right] \nu_s(dz) \right] ds$$
$$\left. + U_J(\tau+\sigma), u_{\tau,x,\alpha(\nu),\nu}(\tau+\sigma) \right\}.$$

Proof. See Proposition 6.1 and ([14], Theorem 8.3.12). We will sketch the proof. Assume there is a $\varphi \in C^1([0,1] \times E)$ such that $U_J - \varphi$ reaches a local maximum at $(t_0, x_0) \in [0,1] \times E$ for which

$$\frac{\partial \varphi}{\partial t}(t_0, x_0) + \min_{\nu \in \mathcal{M}^1_+(Z)} \max_{\mu \in \mathcal{M}^1_+(Y)} \left\{ \int_Z \left[\int_Y J(t_0, x_0, y, z) d\mu(y) \right] d\nu(z) \right.$$
$$\left. + \langle \nabla \varphi(t_0, x_0), \int_Z \left[\int_Y F(t_0, x_0, y, z) d\mu(y) \right] d\nu(z) \rangle \right\} < 0.$$

Hence there exists some $\eta > 0$ such that

$$\frac{\partial \varphi}{\partial t}(t_0, x_0) + \min_{\nu \in \mathcal{M}^1_+(Z)} \max_{\mu \in \mathcal{M}^1_+(Y)} \left\{ \int_Z \left[\int_Y J(t_0, x_0, y, z) d\mu(y) \right] d\nu(z) \right.$$
$$\left. + \langle \nabla \varphi(t_0, x_0), \int_Z \left[\int_Y F(t_0, x_0, y, z) d\mu(y) \right] d\nu(z) \rangle \right\} \leq -\eta < 0.$$

Set

$$\Lambda(t, x, \mu, \nu) = \frac{\partial \varphi}{\partial t}(t, x) + \int_Z \left[\int_Y J(t, x, y, z) d\mu(y) \right] d\nu(z)$$
$$+ \langle \nabla \varphi(t, x), \int_Z \left[\int_Y F(t, x, y, z) d\mu(y) \right] d\nu(z) \rangle.$$

Then we have

$$\min_{\nu \in \mathcal{M}^1_+(Z)} \max_{\mu \in \mathcal{M}^1_+(Y)} \Lambda(t_0, x_0, \mu, \nu) \leq -\eta < 0.$$

Hence there exists some $\overline{\nu} \in \mathcal{M}^1_+(Z)$ such that

$$\max_{\mu \in \mathcal{M}^1_+(Y)} \Lambda(t_0, x_0, \mu, \overline{\nu}) \leq -\eta < 0.$$

Since the mapping

$$(t, x) \mapsto \max_{\mu \in \mathcal{M}^1_+(Y)} \Lambda(t_0, x_0, \mu, \overline{\nu})$$

is continuous there is $\varepsilon > 0$ such that

$$\max_{\mu \in \mathcal{M}^1_+(Y)} \Lambda(t, x, \mu, \overline{\nu}) < -\frac{\eta}{2}$$

for $0 \leq t - t_0 \leq \varepsilon$ and $||x - x_0|| \leq \varepsilon$. As $\dot{u}_{t_0, x_0, \alpha(\nu), \nu}$ is estimated by $||\dot{u}_{t_0, x_0, \alpha(\nu), \nu}(t)|| \leq |K(t)|$ with $|K| \in L^1_{\mathbf{R}}([t_0, 1])$ for all $\nu \in \mathcal{Z}(t_0)$ and for all $\alpha \in \Gamma(t_0)$ $\Gamma(t_0)$ so we can choose $\sigma > 0$ such that $||u_{t_0, x_0, \alpha(\nu), \nu}(t) - u_{t_0, x_0, \alpha(\nu), \nu}(t_0)|| \leq \int_{t_0}^{t_0 + \sigma} |K(t)| dt \leq \varepsilon$ for all $t \in [t_0, t_0 + \sigma]$ and for all

$\nu \in \mathcal{Z}(t_0)$ and for all $\alpha \in \Gamma(t_0)$. Then the constant control $\overline{\nu}_t = \overline{\nu}, \forall t \in [t_0, 1]$ belongs to $\mathcal{Z}(t_0)$ and $\alpha(\overline{\nu})$ belongs to $\mathcal{Y}(t_0)$ for all $\alpha \in \Gamma(t_0)$ so that by integrating we have

$$\int_{t_0}^{t_0+\sigma} \Lambda(t, u_{t_0,x_0,\alpha(\overline{\nu}),\overline{\nu}}(t), \alpha(\overline{\nu})_t, \overline{\nu}_t)dt < -\frac{\sigma\eta}{2}$$

for all $\alpha \in \Gamma(t_0)$. Thus

$$\sup_{\alpha\in\Gamma(t_0)} \Bigg\{\int_{t_0}^{t_0+\sigma} [\int_Z \Big[\int_Y J(t, u_{t_0,x_0,\alpha(\overline{\nu}),\overline{\nu}}(t), y, z)\alpha(\overline{\nu})_t(dy)\Big]\overline{\nu}_t(dz)]dt$$
$$+\langle\nabla\varphi(t, u_{t_0,x_0,\alpha(\overline{\nu}),\overline{\nu}}(t)), \int_Z\Big[\int_Y F(t, u_{t_0,x_0,\alpha(\overline{\nu}),\overline{\nu}}(t), y, z)\alpha(\overline{\nu})_t(dy)\Big]\overline{\nu}_t(dz)\rangle$$
$$+ \frac{\partial\varphi}{\partial t}(t, u_{t_0,x_0,\alpha(\overline{\nu}),\overline{\nu}}(t))\Bigg\} < -\frac{\sigma\eta}{2}. \quad (6.2.3)$$

As U_J satisfies the DPP, we have

$$U_J(t_0, x_0) \le \sup_{\alpha\in\Gamma(t_0)} \Bigg\{\int_{t_0}^{t_0+\sigma} \Big[\int_Z\Big[\int_Y J(t, u_{t_0,x_0,\alpha(\overline{\nu}),\overline{\nu}}(t), y, z)\alpha(\overline{\nu})_t(dy)\Big]\overline{\nu}_t(dz)\Big]dt$$
$$+ U_J(t_0+\sigma, u_{t_0,x_0,\alpha(\overline{\nu}),\overline{\nu}}(t_0+\sigma))\Bigg\}.$$

Hence, for every $n \in \mathbf{N}$, there exists $\alpha^n \in \Gamma(t_0)$ such that

$$U_J(t_0, x_0) \le \int_{t_0}^{t_0+\sigma} \Big[\int_Z\Big[\int_Y J(t, u_{t_0,x_0,\alpha^n(\overline{\nu}),\overline{\nu}}(t), y, z)\alpha^n(\overline{\nu})_t(dy)\Big], \overline{\nu}_t(dz)\Big]dt$$
$$+U_J(t_0+\sigma, u_{t_0,x_0,\alpha^n(\overline{\nu}),\overline{\nu}}(t_0+\sigma)) + \frac{1}{n}. \quad (6.2.4)$$

But $U_J - \varphi$ has a local maximum at (t_0, x_0), for small enough σ

$$U_J(t_0, x_0) - \varphi(t_0, x_0) \ge U_J(t_0+\sigma, u_{t_0,x_0,\alpha(\nu),\nu}(t_0+\sigma))$$
$$-\varphi(t_0+\sigma, u_{t_0,x_0,\alpha(\nu),\nu}(t_0+\sigma)) \quad (6.2.5)$$

for any trajectory solution $u_{t_0,x_0,\alpha(\nu),\nu}$ associated with control $(\alpha(\nu), \nu)$ ($\alpha \in \Gamma(t_0)$, $\nu \in \mathcal{Z}(t_0)$). From (6.2.4) and (6.2.5) we deduce

$$U_J(t_0+\sigma, u_{t_0,x_0,\alpha^n(\overline{\nu}),\overline{\nu}}(t_0+\sigma)) - \varphi(t_0+\sigma, u_{t_0,x_0,\alpha^n(\overline{\nu}),\overline{\nu}}(t_0+\sigma))$$
$$\le \int_{t_0}^{t_0+\sigma} \Big[\int_Z\Big[\int_Y J(t, u_{t_0,x_0,\alpha^n(\overline{\nu}),\overline{\nu}}(t), y, z)\alpha^n(\overline{\nu})_t(dy)\Big]\overline{\nu}_t(dz)\Big]dt$$
$$+ U_J(t_0+\sigma, u_{t_0,x_0,\alpha^n(\overline{\nu}),\overline{\nu}}(t_0+\sigma) + \frac{1}{n} - \varphi(t_0, x_0).$$

Thus we have

$$0 \leq \int_{t_0}^{t_0+\sigma} \left[\int_Z \left[\int_Y J(t, u_{t_0,x_0,\alpha^n(\overline{v}),\overline{v}}(t), y, z)\alpha^n(\overline{v})_t(dy) \right], \overline{v}_t(dz) \right] dt$$
$$+ \varphi(t_0+\sigma, u_{t_0,x_0,\alpha^n(\overline{v}),\overline{v}}(t_0+\sigma)) - \varphi(t_0, x_0) + \frac{1}{n} \tag{6.2.6}$$

But

$$\varphi(t_0+\sigma, u_{t_0,x_0,\alpha^n(\overline{v}),\overline{v}}(t_0+\sigma)) - \varphi(t_0, x_0)$$
$$= \int_{t_0}^{t_0+\sigma} \langle \nabla\varphi(t, u_{t_0,x_0,\alpha^n(\overline{v}),\overline{v}}(t)), \dot{u}_{t_0,x_0,\alpha^n(\overline{v}),\overline{v}}(t) \rangle dt$$
$$+ \int_{t_0}^{t_0+\sigma} \frac{\partial \varphi}{\partial t}(t, u_{t_0,x_0,\alpha^n(\overline{v}),\overline{v}}(t)) dt \tag{6.2.7}$$

and

$$\dot{u}_{t_0,x_0,\alpha^n(\overline{v}),\overline{v}}(t) = \int_Z \left[\int_Y F(t, u_{t_0,x_0,\alpha^n(\overline{v}),\overline{v}}(t), y, z)\alpha^n(\overline{v})_t(dy) \right] \overline{v}_t(dz)$$

so that by combining with (6.2.7)

$$\varphi(t_0+\sigma, u_{t_0,x_0,\alpha^n(\overline{v}),\overline{v}}(t_0+\sigma)) - \varphi(t_0, x_0)$$
$$= \int_{t_0}^{t_0+\sigma} \langle \nabla\varphi(t, u_{t_0,x_0,\alpha^n(\overline{v}),\overline{v}}(t)), \int_Z \left[\int_Y F(t, u_{t_0,x_0,\alpha^n(\overline{v}),\overline{v}}(t), y, z)\alpha^n(\overline{v})_t(dy) \right] \overline{v}_t(dz) \rangle dt$$
$$+ \int_{t_0}^{t_0+\sigma} \frac{\partial \varphi}{\partial t}(t, u_{t_0,x_0,\alpha^n(\overline{v}),\overline{v}}(t)). \tag{6.2.8}$$

From (6.2.6) and (6.2.8) we deduce

$$0 \leq \int_{t_0}^{t_0+\sigma} \left[\int_Z \left[\int_Y J(t, u_{t_0,x_0,\alpha^n(\overline{v}),\overline{v}}(t), y, z)\alpha^n(\overline{v})_t(dy) \right] \overline{v}_t(dz) \right] dt$$
$$\int_{t_0}^{t_0+\sigma} \langle \nabla\varphi(t, u_{t_0,x_0,\alpha^n(\overline{v}),\overline{v}}(t)), \int_Z \left[\int_Y F(t, u_{t_0,x_0,\alpha^n(\overline{v}),\overline{v}}(t), y, z)\alpha^n(\overline{v})_t(dy) \right] \overline{v}_t(dz) \rangle dt$$
$$+ \int_{t_0}^{t_0+\sigma} \frac{\partial \varphi}{\partial t}(t, u_{t_0,x_0,\alpha^n(\overline{v}),\overline{v}}(t)) + \frac{1}{n}. \tag{6.2.9}$$

Using (6.2.3) and (6.2.9) it follows that $0 < \frac{\sigma\eta}{2} < \frac{1}{n}$ for every $n \in \mathbf{N}$. Passing to the limit when n goes to ∞ in the preceding inequality yields a contradiction.

Next assume that $U_J - \varphi$ has a local minimum at $(t_0, x_0) \in [0, 1] \times E$. We must prove that

$$\frac{\partial\varphi}{\partial t}(t_0, x_0) + \min_{\nu\in\mathcal{M}^1_+(Z)} \max_{\nu\in\mathcal{M}^1_+(Z)} \left\{ \int_Z \left[\int_Y J(t_0, x_0, y, z) d\mu(y) \right] d\nu(z) \right.$$
$$\left. + \left\langle \nabla\varphi(t_0, x_0), \int_Z \left[\int_Y F(t_0, x_0, y, z) d\mu(y) \right] d\nu(z) \right\rangle \right\} \leq 0$$

and so will assume the contrary that

$$\frac{\partial\varphi}{\partial t}(t_0, x_0) + \min_{\nu\in\mathcal{M}^1_+(Z)} \max_{\mu\in\mathcal{M}^1_+(Y)} \left\{ \int_Z \left[\int_Y J(t_0, x_0, y, z) d\mu(y) \right] d\nu(z) \right.$$
$$\left. + \left\langle \nabla\varphi(t_0, x_0), \int_Z \left[\int_Y F(t_0, x_0, y, z) d\mu(y) \right] d\nu(z) \right\rangle \right\} > \eta > 0.$$

Arguing as in ([14], Lemma 8.3.11(b)) asserts that there exists for all sufficiently small $\sigma > 0$ some $\alpha \in \Gamma(t_0)$ such that

$$\int_{t_0}^{t_0+\sigma} \left[\int_Z \left[\int_Y J(t, u_{t_0,x_0,\alpha(\nu),\nu}(t), y, z) \alpha(\nu)_t(dy) \right] \nu_t(dz) \right] dt$$
$$+ \int_{t_0}^{t_0+\sigma} \langle \nabla\varphi(t, u_{t_0,x_0,\alpha(\nu),\nu}(t)), \int_Z \left[\int_Y F(t, u_{t_0,x_0,\alpha(\nu)\nu}(t), y, z) \alpha(\nu)_t(dy) \right]$$
$$\times \nu_t(dz) \rangle dt$$
$$+ \int_{t_0}^{t_0+\sigma} \frac{\partial\varphi}{\partial t}(t, u_{t_0,x_0,\alpha(\nu),\nu}(t)) \geq \frac{\sigma\eta}{2} \tag{6.2.10}$$

for all $\nu \in \mathcal{Z}(t_0)$. According to the DPP property we have

$$U_J(t_0, x_0)$$
$$\geq \inf_{\nu\in\mathcal{Z}(t_0)} \left\{ \int_{t_0}^{t_0+\sigma} \left[\int_Z \left[\int_Y J(t, u_{t_0,x_0,\alpha(\nu),\nu}(t), y, z) \alpha(\nu)_t(dy) \right] \nu_t(dz) \right] dt \right.$$
$$\left. + U_J(t_0 + \sigma, u_{t_0,x_0,\alpha(\nu),\nu}(t_0 + \sigma)) \right\}.$$

Hence, for every $n \in \mathbf{N}$, there exists $\nu^n \in \mathcal{Z}(t_0)$ such that

$$U_J(t_0, x_0) \geq \int_{t_0}^{t_0+\sigma} \left[\int_Z \left[\int_Y J(t, u_{t_0,x_0,\alpha(\nu^n),\nu^n}(t), y, z) \alpha(\nu^n)_t(dy) \right] \nu^n_t(dz) \right] dt$$
$$+ U_J(t_0 + \sigma, u_{t_0,x_0,\alpha(\nu^n),\nu^n}(t_0 + \sigma)) - \frac{1}{n}. \tag{6.2.11}$$

But $U_J - \varphi$ has a local minimum at (t_0, x_0), for small enough σ

$$U_J(t_0, x_0) - \varphi(t_0, x_0) \le U_J(t_0 + \sigma, u_{t_0,x_0,\alpha(\nu),\nu}(t_0 + \sigma)) \\ -\varphi(t_0 + \sigma, u_{t_0,x_0,\alpha(\nu),\nu}(t_0 + \sigma)) \quad (6.2.12)$$

for any trajectory solution $u_{t_0,x_0,\alpha(\nu),\nu}$ associated with control $(\alpha(\nu), \nu)$ ($\alpha \in \Gamma(t_0)$, $\nu \in \mathcal{Z}(t_0)$). From (6.2.11) and (6.2.12) we deduce

$$U_J(t_0 + \sigma, u_{t_0,x_0,\alpha(\nu^n),\nu^n}(t_0 + \sigma)) - \varphi(t_0 + \sigma, u_{t_0,x_0,\alpha(\nu^n),\nu^n}(t_0 + \sigma)) \\ \ge \int_{t_0}^{t_0+\sigma} \left[\int_Z \left[\int_Y J(t, u_{t_0,x_0,\alpha(\nu^n),\nu^n}(t), y, z)\alpha(\nu^n)_t(dy)\right] \nu_t^n(dz)\right] dt \\ +U_J(t_0 + \sigma, u_{t_0,x_0,\alpha(\nu^n),\nu^n}(t_0 + \sigma)) - \frac{1}{n} - \varphi(t_0, x_0).$$

Thus we have

$$0 \ge \int_{t_0}^{t_0+\sigma} \left[\int_Z \left[\int_Y J(t, u_{t_0,x_0,\alpha(\nu^n),\nu^n}(t), y, z)\alpha(\nu^n)_t(dy)\right] \nu_t^n(dz)\right] dt \\ +\varphi(t_0 + \sigma, u_{t_0,x_0,\alpha(\nu^n),\nu^n}(t_0 + \sigma)) - \varphi(t_0, x_0) - \frac{1}{n}. \quad (6.2.13)$$

But

$$\varphi(t_0 + \sigma, u_{t_0,x_0,\alpha(\nu^n),\nu^n}(t_0 + \sigma)) - \varphi(t_0, x_0) \\ = \int_{t_0}^{t_0+\sigma} \langle \nabla\varphi(t, u_{t_0,x_0,\alpha(\nu^n),\nu^n}(t)), \dot{u}_{t_0,x_0,\alpha(\nu^n),\nu^n}(t)\rangle dt \\ + \int_{t_0}^{t_0+\sigma} \frac{\partial\varphi}{\partial t}(t, u_{t_0,x_0,\alpha(\nu^n),\nu^n}(t))dt \quad (6.2.14)$$

and

$$\dot{u}_{t_0,x_0,\alpha(\nu^n),\nu^n}(t) = \int_Z \left[\int_Y F(t, u_{t_0,x_0,\alpha(\nu^n),\nu^n}(t), y, z)\alpha(\nu^n)_t(dy)\right] \nu_t^n(dz)$$

so that from (6.2.14)

$$\varphi(t_0 + \sigma, u_{t_0,x_0,\alpha(\nu^n),\nu^n}(t_0 + \sigma)) - \varphi(t_0, x_0) \\ = \int_{t_0}^{t_0+\sigma} \langle \nabla\varphi(t, u_{t_0,x_0,\alpha(\nu^n),\nu^n}(t)), \\ \int_Z \left[\int_Y F(t, u_{t_0,x_0,\alpha(\nu^n),\nu^n}(t), y, z)\alpha(\nu^n)_t(dy)\right] \nu_t^n(dz)\rangle dt \\ + \int_{t_0}^{t_0+\sigma} \frac{\partial\varphi}{\partial t}(t, u_{t_0,x_0,\alpha(\nu^n),\nu^n}(t)). \quad (6.2.15)$$

From (6.2.13) and (6.2.15) we deduce

$$0 \geq \int_{t_0}^{t_0+\sigma} \left[\int_Z \left[\int_Y J(t, u_{t_0,x_0,\alpha(\nu^n),\nu^n}(t), y, z)\alpha(\nu^n)_t(dy) \right] \nu_t^n(dz) \right] dt$$
$$+ \int_{t_0}^{t_0+\sigma} \langle \nabla\varphi(t, u_{t_0,x_0,\alpha(\nu^n),\nu^n}(t)),$$
$$\int_Z \left[\int_Y F(t, u_{t_0,x_0,\alpha(\nu^n),\nu^n}(t), y, z)\alpha(\nu^n)_t(dy) \right] \nu_t^n(dz)\rangle dt$$
$$+ \int_{t_0}^{t_0+\sigma} \frac{\partial\varphi}{\partial t}(t, u_{t_0,x_0,\alpha(\nu^n),\nu^n}(t)) - \frac{1}{n}. \tag{6.2.16}$$

Using (6.2.10) and (6.2.16) it follows that $\frac{1}{n} \geq \frac{\sigma\eta}{2} > 0$ for every $n \in \mathbf{N}$. Passing to the limit when n goes to ∞ in the preceding inequality yields a contradiction. □

Taking account into the sweeping process introduced by J.J. Moreau [26] and its modelisation in Mathematical Economics [25], we finish the paper with an application to the DPP and viscosity property for the value function associated with a sweeping process. Compare with Theorem 3.5 in [17] dealing with sweeping process involving Young measure control and Theorem 4.2 dealing with (SODE). Here E is a separable Hilbert space.

Proposition 6.3. *Let $C : [0, T] \to ck(E)$ be a convex compact valued L-Lipschitzean mapping:*

$$|d(x, C(t)) - d(y, C(\tau))| \leq L|t - \tau| + ||x - y||, \forall x, y \in E \times E, \forall t,$$
$$\tau \in [0, T] \times [0, T].$$

Let Z be a convex compact subset in E and $\mathcal{S}_Z^1$ is the set of all integrable mappings $f : [0, T] \to Z$. Assume that $J : [0, T] \times E \times E \to \mathbf{R}$ is bounded and continuous such that $J(t, x, .)$ is convex for every $(t, x) \in [0, T] \times E$. Let us consider the value function

$$V_J(\tau, x) = \sup_{f \in \mathcal{S}_Z^1} \left\{ \int_\tau^T J(t, u_{\tau,x,f}(t), f(t))\, dt \right\}, \; (\tau, x) \in [0, T] \times E$$

where $u_{\tau,x,f}$ is the trajectory solution on $[\tau, T]$ associated the control $f \in \mathcal{S}_Z^1$ starting from $x \in E$ at time τ to the sweeping process $(\mathcal{PSW})(C; f; x)$

$$\begin{cases} -\dot{u}_{\tau,x,f}(t) - f(t) \in N_{C(t)}(u_{\tau,x,f}(t)), \; t \in [\tau, T] \\ u_{\tau,x,f}(\tau) = x \end{cases}$$

and the Hamiltonian

$$H(t,x,\rho) = \sup_{z\in Z}\{-\langle \rho, z\rangle + J(t,x,z)\} + \delta^*(\rho, -M\partial[d_{C(t)}](x))$$

where $M := L + 2|Z|$, $(t,x,\rho) \in [0,T] \times E \times E$ *and* $\partial[d_{C(t)}](x)$ *denotes the subdifferential of the distance functions* $x \mapsto d_{C(t)}x$. *Then* V_J *has the DPP property*

$$V_J(\tau,x) = \sup_{f\in S^1_Z}\left\{\int_\tau^{\tau+\sigma} J(t,u_{\tau,x,f}(t),f(t))dt + V_J(\tau+\sigma, u_{\tau,x,f}(\tau+\sigma))\right\}$$

and is a viscosity subsolution to the HJB equation

$$\frac{\partial U}{\partial t}(t,x) + H(t,x,\nabla U(t,x)) = 0$$

that is, for any $\varphi \in C^1([0,T]) \times E)$ *for which* $V_J - \varphi$ *reaches a local maximum at* $(t_0,x_0) \in [0,T] \times E$, *we have*

$$H(t_0,x_0,\nabla\varphi(t_0,x_0)) + \frac{\partial\varphi}{\partial t}(t_0,x_0) \geq 0.$$

Proof. We prove first that V_J has the DPP property by applying the continuous property of the solution with respect to the state and the control (see Lemma 6.1 below) and lower semicontinuity of the integral functional ([14], Theorem 8.1.6). We omit the proof of Lemma 6.1 because it is an adaptation of the proof of Lemma 4.1 in [17].

Lemma 6.1. *Let* u_{τ,x^n,f^n} *be the trajectory solution on* $[\tau,T]$ *associated the control* $f^n \in S^1_Z$ *starting from* $x^n \in E$ *at time* τ *to the sweeping process* $(\mathcal{PSW})(C; f^n; x)$

$$\begin{cases} -\dot u_{\tau,x^n,f^n}(t) - f^n(t) \in N_{C(t)}(u_{\tau,x^n,f^n}(t)) \\ u_{\tau,x^n,f^n}(\tau) = x^n \in C(\tau) \end{cases}$$

(a) If (x^n) *converges to* x^∞ *and* f^n *converges* $\sigma(L^1_E, L^\infty_E)$ *to* f^∞, *then* u_{τ,x^n,f^n} *converges uniformly to* $u_{\tau,x^\infty,f^\infty}$, *which is the Lipschitz solution of the sweeping process* $(\mathcal{PSW})(C; f^\infty; x^\infty)$

$$\begin{cases} -\dot u_{\tau,x^\infty,f^\infty}(t) - f^\infty(t) \in N_{C(t)}(u_{\tau,x^\infty,f^\infty}(t)) \\ u_{\tau,x^\infty,f^\infty}(\tau) = x^\infty \in C(\tau) \end{cases}$$

(b) Let $J : [0,1] \times (E \times E) \to]-\infty, +\infty]$ *be a normal integrand such that* $J(t,x,.)$ *is convex on* E *for all* $(t,x) \in [0,T] \times E$ *and that*

$$J(t, u_{\tau,x^n,f^n}(t), f^n(t)) \geq \beta_n(t)$$

for all $n \in \mathbf{N}$ *and for all* $t \in [0,T]$ *for some uniformly integrable sequence* $(\beta_n)_{n \in \mathbf{N}}$ *in* $L^1_{\mathbf{R}}([0,T])$*, then we have*

$$\liminf_{n\to\infty} \int_\tau^T J(t, u_{\tau,x^n,f^n}(t), f^n(t))\,dt \geq \int_\tau^T J(t, u_{\tau,x^\infty,f^\infty}(t), f^\infty(t))\,dt.$$

Let us focus on the expression of $V_J(\tau+\sigma, u_{\tau,x,f}(\tau+\sigma))$

$$V_J(\tau+\sigma, u_{\tau,x,f}(\tau+\sigma)) = \sup_{g \in S^1_Z} \left\{ \int_{\tau+\sigma}^T J(t, v_{\tau+\sigma, u_{\tau,x,f}(\tau+\sigma),g}(t), g(t))dt \right\}$$

where $v_{\tau+\sigma, u_{\tau,x,f}(\tau+\sigma),g}$ denotes the trajectory solution on $[\tau+\sigma, T]$ associated with the control $g \in S^1_Z$ starting from $u_{\tau,x,f}(\tau+\sigma)$ at time $\tau+\sigma$.

Main Fact: $f \to V_J(\tau+\sigma, u_{\tau,x,f}(\tau+\sigma))$ is lower semicontinuous on S^1_Z (endowed with the $\sigma(L^1_E, L^\infty_E)$-topology). Let $(f_n, g_n) \in \mathcal{S}^1_Z \times \mathcal{S}^1_Z$ such that $f_n \to f \in \mathcal{S}^1_Z$ and $g_n \to g \in \mathcal{S}^1_Z$. By Lemma 6.1, u_{τ,x,f_n} converges uniformly to $u_{\tau,x,f}$ and $v_{\tau+\sigma,u_{\tau,x,f_n}(\tau+\sigma),g_n}$ converges uniformly to $v_{\tau+\sigma,u_{\tau,x,f}(\tau+\sigma),g}$ so that by invoking the lower semicontinuity of integral functional ([14], Theorem 8.1.6) we get

$$\liminf_{n\to\infty} \int_\tau^{\tau+\sigma} J(t, u_{\tau,x,f_n}(t), f_n(t))dt \geq \int_\tau^{\tau+\sigma} J(t, u_{\tau,x,f}(t), f(t))dt$$

- $$\liminf_{n\to\infty} \int_{\tau+\sigma}^T J(t, v_{\tau+\sigma,u_{\tau,x,f_n}(\tau+\sigma),g_n}(t), g_n(t))dt \geq \int_{\tau+\sigma}^T J(t, v_{\tau+\sigma,u_{\tau,x,f}(\tau+\sigma),g}(t), g(t))dt$$

proving that the mapping $f \mapsto \int_\tau^{\tau+\sigma} J(t, u_{\tau,x,f}(t), f(t))dt$ is lower semicontinuous on $\mathcal{S}^1_Z$ and the mapping $(f,g) \mapsto \int_{\tau+\sigma}^T J(t, v_{\tau+\sigma,u_{\tau,x,f}(\tau+\sigma),g}(t), g(t))dt$ is lower semicontinuous on $\mathcal{S}^1_Z \times \mathcal{S}^1_Z$. It follows that the mapping $f \mapsto V_J(\tau+\sigma, u_{\tau,x,f}(\tau+\sigma))$ is lower semicontinuous on $\mathcal{S}^1_Z$ and so is the mapping $f \mapsto \int_\tau^{\tau+\sigma} J(t, u_{\tau,x,f}(t), f(t))dt + V_J(\tau+\sigma, u_{\tau,x,f}(\tau+\sigma))$. Now the DPP property for V_J follows the same line of the proof of Theorem 4.1. This fact allows to obtain the required viscosity property. Let us recall the following

Lemma 6.2. *Let $(t_0, x_0) \in [0, T] \times E$ and let Z be a convex compact subset in E. Let $\Lambda : [0, T] \times E \times Z \to \mathbf{R}$ be an upper semicontinuous function such that the restriction of Λ to $[0, T] \times B \times Z$ is bounded on any bounded subset B of E. If*

$$max_{z \in Z} \Lambda(t_0, x_0, z) < -\eta < 0$$

for some $\eta > 0$, then there exists $\sigma > 0$ such that

$$\sup_{f \in \mathcal{S}_Z^1} \int_{t_0}^{t_0+\sigma} \Lambda(t, u_{t_0,x_0,f}(t), f(t))dt < -\frac{\sigma\eta}{2}$$

where $u_{t_0,x_0,f}$ is the trajectory solution associated with the control $f \in \mathcal{S}_Z^1$ starting from x_0 at time t_0 to

$$\begin{cases} -\dot{u}_{t_0,x_0,f}(t) - f(t) \in N_{C(t)}(u_{t_0,x_0,f}(t)),\ t \in [t_0, T] \\ u_{t_0,x_0,f}(t_0) = x_0. \end{cases}$$

Assume by contradiction that there exists a $\varphi \in C^1([0, T] \times E)$ and a point $(t_0, x_0) \in [0, T] \times E$ for which

$$\frac{\partial \varphi}{\partial t}(t_0, x_0) + H(t_0, x_0, \nabla\varphi(t_0, x_0)) \le -\eta < 0 \quad \text{for} \quad \eta > 0.$$

Applying Lemma 6.2 by taking

$$\begin{aligned} \Lambda(t, x, z) = J(t, x, z) - \langle \nabla\varphi(t, x), z\rangle + \delta^*(\nabla\varphi(t, x), \\ - M\,\partial[d_{C(t)}](x)) + \frac{\partial\varphi}{\partial t}(t, x) \end{aligned}$$

provides some $\sigma > 0$ such that

$$\begin{aligned} \sup_{f \in \mathcal{S}_Z^1} \bigg\{ \int_{t_0}^{t_0+\sigma} J(t, u_{t_0,x_0,f}(t), f(t))\,dt - \int_{t_0}^{t_0+\sigma} \langle \nabla\varphi(t, u_{t_0,x_0,f}(t)), f(t)\rangle\,dt \\ + \int_{t_0}^{t_0+\sigma} \delta^*(\nabla\varphi(t, u_{t_0,x_0,f}(t)), -M\,\partial[d_{C(t)}](u_{t_0,x_0,f}(t)))\,dt \\ + \int_{t_0}^{t_0+\sigma} \frac{\partial\varphi}{\partial t}(t, u_{t_0,x_0,f}(t))\,dt \bigg\} < -\frac{\sigma\eta}{2} \end{aligned} \tag{6.3.1}$$

where $u_{t_0,x_0,f}$ is the trajectory solution associated with the control $f \in \mathcal{S}_Z^1$ starting from x_0 at time t_0 to the sweeping process $(\mathcal{PSW})(C; f; x)$

$$\begin{cases} -\dot{u}_{t_0,x_0,f}(t) - f(t) \in N_{C(t)}(u_{t_0,x_0,f}(t)),\ t \in [t_0, T] \\ u_{t_0,x_0,f}(t_0) = x_0. \end{cases}$$

Applying the dynamic programming principle gives

$$V_J(t_0, x_0) = \sup_{f\in\mathcal{S}^1_Z} \left[\int_{t_0}^{t_0+\sigma} J(t, u_{t_0,x_0,f}(t), f(t))dt \right.$$
$$\left. + V_J(t_0+\sigma, u_{t_0,x_0,f}(\tau+\sigma))\right]. \tag{6.3.2}$$

Since $V_J - \varphi$ has a local maximum at (t_0, x_0), for small enough σ

$$V_J(t_0, x_0) - \varphi(t_0, x_0) \geq V_J(t_0+\sigma, u_{t_0,x_0,f}(t_0+\sigma))$$
$$-\varphi(t_0+\sigma, u_{t_0,x_0,f}(t_0+\sigma)) \tag{6.3.3}$$

for all $f \in \mathcal{S}^1_Z$. For each $n \in \mathbf{N}$, there exists $f^n \in \mathcal{S}^1_Z$ such that

$$V_J(t_0, x_0) \leq \int_{t_0}^{t_0+\sigma} J(t, u_{t_0,x_0,f^n}(t), f^n(t))\, dt + V_J(t_0+\sigma, u_{t_0,x_0,f^n}(t_0+\sigma)) + \frac{1}{n}. \tag{6.3.4}$$

From (6.3.3) and (6.3.4) we deduce that

$$V_J(t_0+\sigma, u_{t_0,x_0,f^n}(t_0+\sigma)) - \varphi(t_0+\sigma, u_{t_0,x_0,f^n}(t_0+\sigma))$$
$$\leq \int_{t_0}^{t_0+\sigma} J(t, u_{t_0,x_0,f^n}(t), f^n(t))dt + \frac{1}{n}$$
$$-\varphi(t_0, x_0) + V_J(t_0+\sigma, u_{t_0,x_0,f^n}(t_0+\sigma)).$$

Therefore we have

$$0 \leq \int_{t_0}^{t_0+\sigma} J(t, u_{t_0,x_0,f^n}(t), f^n(t))\, dt + \varphi(t_0+\sigma, u_{t_0,x_0,f^n}(t_0+\sigma) - \varphi(t_0, x_0) + \frac{1}{n}. \tag{6.3.5}$$

As $\varphi \in C^1([0, T] \times E)$ we have

$$\varphi(t_0+\sigma, u_{t_0,x_0,f^n}(t_0+\sigma)) - \varphi(t_0, x_0)$$
$$= \int_{t_0}^{t_0+\sigma} \langle \nabla\varphi(t, u_{t_0,x_0,f^n}(t)), \dot{u}_{t_0,x_0,f^n}(t)\rangle\, dt + \int_{t_0}^{t_0+\sigma} \frac{\partial\varphi}{\partial t}(t, u_{t_0,x_0,f^n}(t))\, dt. \tag{6.3.6}$$

Since u_{t_0,x_0,f^n} is the trajectory solution starting from x_0 at time t_0 to the sweeping process $(\mathcal{PSW})(C; f^n; x)$

$$\begin{cases} -\dot{u}_{t_0,x_0,f^n}(t) - f^n(t) \in N_{C(t)}(u_{t_0,x_0,f^n}(t)),\ t \in [t_0, T] \\ u_{t_0,x_0,f^n}(t_0) = x_0 \end{cases}$$

by the classical property of normal convex cone and the estimation $||\dot{u}_{t_0,x_0,f^n}(t) - f^n(t)|| \le L + 2|Z| = M$ we get

$$-\dot{u}_{t_0,x_0,f^n}(t) - f^n(t) \in M\, \partial[d_{C(t)}](u_{t_0,x_0,f^n}(t))$$

so that (6.3.6) yields

$$\begin{aligned}
&\varphi(t_0+\sigma, u_{t_0,x_0,f^n}(t_0+\sigma)) - \varphi(t_0, x_0)\\
&= \int_{t_0}^{t_0+\sigma} \langle \nabla\varphi(t, u_{t_0,x_0,f^n}(t)), \dot{u}_{t_0,x_0,f^n}(t)\rangle\, dt\\
&+ \int_{t_0}^{t_0+\sigma} \frac{\partial\varphi}{\partial t}(t, u_{t_0,x_0,f^n}(t))\, dt\\
&\le -\int_{t_0}^{t_0+\sigma} \langle \nabla\varphi(t, u_{t_0,x_0,f^n}(t)), f^n(t)\rangle\, dt\\
&+ \int_{t_0}^{t_0+\sigma} \delta^*(\nabla\varphi(t, u_{t_0,x_0,f^n}(t)), -M\,\partial[d_{C(t)}](u_{t_0,x_0,f^n}(t)))\, dt\\
&+ \int_{t_0}^{t_0+\sigma} \frac{\partial\varphi}{\partial t}(t, u_{t_0,x_0,f^n}(t))\, dt. \qquad (6.3.7)
\end{aligned}$$

Putting the estimate (6.3.7) in (6.3.5) we get

$$\begin{aligned}
0 \le &\int_{t_0}^{t_0+\sigma} J(t, u_{t_0,x_0,f^n}(t), f^n(t))dt - \int_{t_0}^{t_0+\sigma} \langle \nabla\varphi(t, u_{t_0,x_0,f^n}(t)), f^n(t)\rangle\, dt\\
&+ \int_{t_0}^{t_0+\sigma} \delta^*(\nabla\varphi(t, u_{t_0,x_0,f^n}(t)), -M\partial[d_{C(t)}](u_{t_0,x_0,f^n}(t)))\, dt\\
&+ \int_{t_0}^{t_0+\sigma} \frac{\partial\varphi}{\partial t}(t, u_{t_0,x_0,f^n}(t))\, dt + \frac{1}{n} \qquad (6.3.8)
\end{aligned}$$

so that (6.3.1) and (6.3.8) give $0 < \frac{\sigma\eta}{2} < \frac{1}{n}$ for all $n \in \mathbf{N}$. Passing to the limit when n goes ∞ in this inequality gives a contradiction. □

Viscosity problem governed by sweeping process with strategies and Young measures

$$\begin{aligned}
&\dot{u}_{\tau,x,\alpha(\nu),\nu}(t) \in \int_Z \left[\int_Y F(t, u_{\tau,x,\alpha(\nu),\nu}(t), y, z)\alpha(\nu)_t(dy)\right] \nu_t(dz)\\
&- N_{C(t)}(u_{\tau,x,\alpha(\nu),\nu}(t)), t \in [\tau, 1],\\
&u_{\tau,x,\alpha(\nu),\nu}(\tau) = x \in C(\tau),\\
&U_J(\tau, x) = \sup_{\alpha\in\Gamma(\tau)} \inf_{\nu\in\mathcal{Z}(\tau)} \left\{\int_\tau^1 \left[\int_Z \left[\int_Y J(t, u_{\tau,x,\alpha(\nu),\nu}(t), y, z)\alpha(\nu)_t(dy)\right] \nu_t(dz)\right] dt\right\}
\end{aligned}$$

where the integrand J, the upper value function U_J, the data Y, Z and F are defined as in Proposition 6.2, is an open problem. Further related results dealing with continuous and bounded variation (BVC) solution in sweeping process governed by non empty interior closed convex valued continuous mappings are available in [17, 18].

References

1. Amrani, A., Castaing, C., Valadier, M.: Convergence in Pettis norm under extreme points condition. Vietnam J. Math. **26**(4), 323–335 (1998)
2. Azzam, D.L., Castaing, C., Thibault, L.: Three point boundary value problems for second order differential inclusions in Banach spaces. Control Cybern. **31**, 659–693 (2002). Well-posedness in optimization and related topics (Warsaw, 2001)
3. Azzam, D.L., Makhlouf, A., Thibault, L.: Existence and relaxation theorem for a second order differential inclusion. Numer. Funct. Anal. Optim. **31**, 1103–1119 (2010)
4. Bardi, M., Capuzzo Dolcetta, I.: Optimal Control and Viscosity Solutions of Hamilton-Jacobi-Bellman Equations. Birkhauser, Boston (1997)
5. Castaing, C.: Topologie de la convergence uniforme sur les parties uniformément intégrables de L^1_E et théorème de compacité faible dans certains espaces du type Köthe-Orlicz. Sém. Anal. Convexe **10**, 5.1–5.27 (1980)
6. Castaing, C.: Weak compactness and convergences in Bochner and Pettis integration. Vietnam J. Math. **24**(3), 241–286 (1996)
7. Castaing, C., Marcellin, S.: Evolution inclusions with pln functions and application to viscosity and controls. J. Nonlinear Convex Anal. **8**(2), 227–255 (2007)
8. Castaing, C., Raynaud de Fitte, P.: On the fiber product of Young measures with applications to a control problem with measures. Adv. Math. Econ. **6**, 1–38 (2004)
9. Castaing, C., Truong, L.X.: Second order differential inclusions with m-points boundary condition. J. Nonlinear Convex Anal. **12**(2), 199–224 (2011)
10. Castaing, C., Truong, L.X.: Some topological properties of solutions set in a second order inclusion with m-point boundary condition. Set Valued Var. Anal. **20**, 249–277 (2012)
11. Castaing, C., Truong, L.X.: Bolza, relaxation and viscosity problems governed by a second order differential equation. J. Nonlinear Convex Anal. **14**(2), 451–482 (2013)

12. Castaing, C., Valadier, M.: Convex analysis and measurable multifunctions. In: Lecture Notes in Mathematics, vol. 580. Springer, Berlin (1977)
13. Castaing, C., Jofre, A., Salvadori, A.: Control problems governed by functional evolution inclusions with Young measures. J. Nonlinear Convex Anal. **5**(1), 131–152 (2004)
14. Castaing, C., Raynaud de Fitte, P., Valadier, M.: Young Measures on Topological Spaces. With Applications in Control Theory and Probability Theory. Kluwer Academic, Dordrecht (2004)
15. Castaing, C., Jofre, A., Syam, A.: Some limit results for integrands and Hamiltonians with application to viscosity. J. Nonlinear Convex Anal. **6**(3), 465–485 (2005)
16. Castaing, C., Raynaud de Fitte, P., Salvadori, A.: Some variational convergence results with application to evolution inclusions. Adv. Math. Econ. **8**, 33–73 (2006)
17. Castaing, C., Monteiro Marquès, M.D.P., Raynaud de Fitte, P.: On a optimal control problem governed by the sweeping process (2013). Preprint
18. Castaing, C., Monteiro Marquès, M.D.P., Raynaud de Fitte, P.: On a Skorohod problem (2013). Preprint
19. El Amri, K., Hess, C.: On the Pettis integral of closed valued multifunction. Set Valued Anal. **8**, 329–360 (2000)
20. Elliot, R.J.: Viscosity Solutions and Optimal Control. Pitman, London (1977)
21. Elliot, R.J., Kalton, N.J.: Cauchy problems for certains Isaacs-Bellman equations and games of survival. Trans. Am. Soc. **198**, 45–72 (1974)
22. Evans, L.C., Souganides, P.E.: Differential games and representation formulas for solutions of Hamilton-Jacobi-Issacs equations. Indiana Univ. Math. J. **33**, 773–797 (1984)
23. Godet-Thobie, C., Satco, B.: Decomposability and uniform integrability in Pettis integration. Questiones Math. **29**, 39–58 (2006)
24. Grothendieck, A.: Espaces vectoriels topologiques. Publicação da Sociedade de Matemática de Sao Paulo (1964)
25. Henry, C.: An existence theorem for a class of differential equation with multivalued right-hand side. J. Math. Anal. Appl. **41**, 179–186 (1973)
26. Moreau, J.J.: Evolution problem associated with a moving set in Hilbert space. J. Differ. Equ. **26**, 347–374 (1977)
27. Satco, B.: Second order three boundary value problem in Banach spaces via Henstock and Henstock-Kurzweil-Pettis integral. J. Math. Appl. **332**, 912–933 (2007)
28. Valadier, M.: Some bang-bang theorems. In: Multifunctions and Integrands, Stochastics Analysis, Approximations and Optimization Proceedings, Catania, 1983. Lecture Notes in Mathematics, vol. 1091, pp. 225–234

Adv. Math. Econ. 18, 61–99 (2014)

Advances in
MATHEMATICAL
ECONOMICS

Stochastic Mesh Methods for Hörmander Type Diffusion Processes

Shigeo Kusuoka[1] **and Yusuke Morimoto**[1,2]

[1] Graduate School of Mathematical Sciences, The University of Tokyo, 3-8-1 Komaba, Meguro-ku, Tokyo 153-8914, Japan
(e-mail: kusuoka@ms.u-tokyo.ac.jp)
[2] Bank of Tokyo Mitsubishi UFJ, Tokyo, Japan

Received: August 29, 2013
Revised: November 15, 2013

JEL classification: C63, G12

Mathematics Subject Classification (2010): 65C05, 60G40

Abstract. In the present paper the authors discuss the efficiency of stochastic mesh methods introduced by Broadie and Glasserman (J Comput Finance 7(4):35–72, 2004). The authors apply stochastic mesh methods to certain type of Hörmander type diffusion processes and show the following. (1) If one carefully takes partitions, the estimated price of American option converges to the real price with probability one. (2) One can obtain better estimates by re-simulation methods discussed in Belomestny (Finance Stoch 15:655–683, 2011), although the order is not so sharp as his result.

Key words: Computational finance, Malliavin calculus, Option pricing, Stochastic mesh method

1. Introduction

Stochastic mesh methods were introduced by Broadie and Glasserman [4], and Avramidis and Hyden [1] and Avramidis and Matzinger [2] proved the efficiency of them in some cases (see [5] also). Also, Belomestny [3] showed in Bermuda options that once we have estimated functions for the so-called continuation values, we have a better estimated value if we construct a pre-optimal stopping time by using these estimated functions and estimate the expectation of pay-off functionals based on this stopping time by re-simulation.

In the present paper, we consider the efficiency of stochastic mesh methods and re-simulation in the case that we apply them to Hörmander type diffusion processes.

Let $N, d \geqq 1$. Let $W_0 = \{w \in C([0,\infty); \mathbf{R}^d);\ w(0) = 0\}$, $\mathcal{F}$ be the Borel algebra over W_0 and μ be the Wiener measure on $(W_0, \mathcal{F})$. Let $B^i : [0,\infty) \times W_0 \to \mathbf{R}$, $i = 1, \ldots, d$, be given by $B^i(t, w) = w^i(t)$, $(t, w) \in [0,\infty) \times W_0$. Then $\{(B^1(t), \ldots, B^d(t); t \in [0,\infty)\}$ is a d-dimensional Brownian motion. Let $B^0(t) = t$, $t \in [0,\infty)$. Let $V_0, V_1, \ldots, V_d \in C_b^\infty(\mathbf{R}^N; \mathbf{R}^N)$. Here $C_b^\infty(\mathbf{R}^N; \mathbf{R}^n)$ denotes the space of $\mathbf{R}^n$-valued smooth functions defined in $\mathbf{R}^N$ whose derivatives of any order are bounded. We regard elements in $C_b^\infty(\mathbf{R}^N; \mathbf{R}^N)$ as vector fields on $\mathbf{R}^N$.

Now let $X(t, x)$, $t \in [0,\infty)$, $x \in \mathbf{R}^N$, be the solution to the Stratonovich stochastic integral equation

$$X(t, x) = x + \sum_{i=0}^{d} \int_0^t V_i(X(s, x)) \circ dB^i(s). \tag{1}$$

Then there is a unique solution to this equation. Moreover we may assume that $X(t, x)$ is continuous in t and smooth in x and $X(t, \cdot) : \mathbf{R}^N \to \mathbf{R}^N$, $t \in [0,\infty)$, is a diffeomorphism with probability one.

Let $\mathcal{A} = \{\emptyset\} \cup \bigcup_{k=1}^{\infty} \{0, 1, \ldots, d\}^k$ and for $\alpha \in \mathcal{A}$, let $|\alpha| = 0$ if $\alpha = \emptyset$, let $|\alpha| = k$ if $\alpha = (\alpha^1, \ldots, \alpha^k) \in \{0, 1, \ldots, d\}^k$, and let $\| \alpha \| = |\alpha| + \mathrm{card}\{1 \leqq i \leqq |\alpha|;\ \alpha^i = 0\}$. Let $\mathcal{A}^*$ and $\mathcal{A}^{**}$ denote $\mathcal{A} \setminus \{\emptyset\}$ and $\mathcal{A} \setminus \{\emptyset, 0\}$, respectively. Also, for each $m \geqq 1$, $\mathcal{A}^{**}_{\leqq m}$, $\{\alpha \in \mathcal{A}^{**};\ \| \alpha \| \leqq m\}$.

We define vector fields $V_{[\alpha]}$, $\alpha \in \mathcal{A}$, inductively by

$$V_{[\emptyset]} = 0, \qquad V_{[i]} = V_i, \quad i = 0, 1, \ldots, d,$$

$$V_{[\alpha * i]} = [V_{[\alpha]}, V_i], \qquad i = 0, 1, \ldots, d.$$

Here $\alpha * i = (\alpha^1, \ldots, \alpha^k, i)$ for $\alpha = (\alpha^1, \ldots, \alpha^k)$ and $i = 0, 1, \ldots, d$.

We say that a system $\{V_i; i = 0, 1, \ldots, d\}$ of vector fields satisfies the following condition (UFG).

(UFG) There are an integer ℓ_0 and $\varphi_{\alpha,\beta} \in C_b^\infty(\mathbf{R}^N)$, $\alpha \in \mathcal{A}^{**}$, $\beta \in \mathcal{A}^{**}_{\leqq \ell_0}$, satisfying the following.

$$V_{[\alpha]} = \sum_{\beta \in \mathcal{A}^{**}_{\leqq \ell_0}} \varphi_{\alpha,\beta} V_{[\beta]}, \qquad \alpha \in \mathcal{A}^{**}.$$

Let $A(x) = (A^{ij}(x))_{i,j=1,\ldots,N}$, $t > 0$ $x \in \mathbf{R}^N$ be a $N \times N$ symmetric matrix given by

$$A^{ij}(x) = \sum_{\alpha \in \mathcal{A}^{**}_{\leqq \ell_0}} V^i_{[\alpha]}(x) V^j_{[\alpha]}(x), \qquad i, j = 1, \ldots, N.$$

Let $h(x) = \det A(x)$, $x \in \mathbf{R}^N$ and $E = \{x \in \mathbf{R}^N;\ h(x) > 0\}$. By Kusuoka–Stroock [7], we see that if $x \in E$, the distribution law of $X(t, x)$ under μ has a smooth density function $p(t, x, \cdot) : \mathbf{R}^N \to [0, \infty)$ for $t > 0$. Moreover, we will show in that $\int_E p(t, x, y)dy = 1$, $x \in E$.

Now let $x_0 \in E$ and fix it throughout this paper. Let $(\Omega, \mathcal{F}, P)$ be a probability space, and $X_\ell : [0, \infty) \times \Omega \to \mathbf{R}^N$, $\ell = 1, 2, \ldots$, be continuous stochastic processes such that the probability laws on $C([0, \infty); \mathbf{R}^N)$ of $X_\ell(\cdot)$ under P and of $X(\cdot, x_0)$ under μ are the same for all $\ell = 1, 2, \ldots$, and that $\sigma\{X_\ell(t);\ t \geqq 0\}$, $\ell = 1, 2, \ldots$, are independent.

Let $q^{(L)}_{s,t} : E \times \Omega \to [0, \infty)$, $t > s \geqq 0$, $L \geqq 1$, be given by

$$q^{(L)}_{s,t}(y, \omega) = \frac{1}{L} \sum_{\ell=1}^{L} p(t - s, X_\ell(s, \omega), y), \qquad y \in E,\ \omega \in \Omega.$$

Let $m(E)$ denote the space of measurable functions on E.

We define a random linear operator $Q^{(L)}_{s,t}$, $t > s \geqq 0$, $L \geqq 1$, defined in $m(E)$ by

$$(Q^{(L)}_{s,t} f)(x) = \frac{1}{L} \sum_{\ell=1}^{L} \frac{p(t - s, x, X_\ell(t)) f(X_\ell(t))}{q^{(L)}_{s,t}(X_\ell(t))}, \qquad x \in E,\ f \in m(E).$$

Now let $T > 0$, and $g : [0, T] \times \mathbf{R}^N \to \mathbf{R}$ be a continuous function with $\sup\{(1 + |x|)^{-1} |g(t, x)|;\ x \in \mathbf{R}^N, t \in [0, T]\} < \infty$. For any $n \geqq 1$, and $0 = t_0 < t_1 < \ldots < t_n = T$, we define $c_{t_k, t_{k+1}, \ldots, t_n} : E \to \mathbf{R}$, and $\tilde{c}^{(L)}_{t_k, t_{k+1}, \ldots, t_n} : E \times \Omega \to \mathbf{R}$, $k = n, n-1, \ldots, 0$, $L \geqq 1$, inductively by $c_{t_n}(x) = \tilde{c}^{(L)}_{t_n}(x) = g(T, x)$, $x \in E$, and

$$c_{t_k, t_{k+1}, \ldots, t_n}(x) = \int_E p(t_{k+1} - t_k, x, y)(g(t_{k+1}, y) \vee c_{t_{k+1}, \ldots, t_n}(y))dy,$$

and

$$\tilde{c}^{(L)}_{t_k, t_{k+1}, \ldots, t_n}(x) = Q^{(L)}_{t_k, t_{k+1}}(g(t_{k+1}, \cdot) \vee \tilde{c}^{(L)}_{t_{k+1}, \ldots, t_n}(\cdot))(x)$$

for $x \in E$ and $k = n - 1, \ldots, 0$.

Then we will show the following.

Theorem 1. *Suppose that* $n(L) \geqq 1$, $0 = t_0^{(L)} < t_1^{(L)} < \ldots < t_{n(L)}^{(L)} = T$. *If there is an* $\varepsilon > 0$ *such that*

$$L^{-(1-\varepsilon)/2} \sum_{k=1}^{n(L)} (t_k^{(L)} - t_{k-1}^{(L)})^{-(N+1)\ell_0/4} \to 0,$$

then

$$E[|\tilde{c}^{(L)}_{t_0^{(L)}, t_1^{(L)}, \ldots, t_{n(L)}^{(L)}}(x_0) - c_{t_0^{(L)}, t_1^{(L)}, \ldots, t_{n(L)}^{(L)}}(x_0)|^2] \to 0, \qquad L \to \infty.$$

Let $n \geqq 1$, and $0 = T_0 < T_1 < \ldots < T_n = T$ and fix them. For each $\omega \in \Omega$, let $\hat{\tau}_{L,\omega} W_0 \to \{T_1, \ldots, T_n\}$ be a stopping time given by

$$\hat{\tau}_{L,\omega} = \min\{T_k;\ k = 1, 2, \ldots, n,\ \tilde{c}^L_{T_k, T_{k+1}, \ldots, T_n}(X(T_k, x_0), \omega) \leqq g(T_k, X(T_k, x_0))\}.$$

Let $\hat{c} : \Omega \to \mathbf{R}$ be given by

$$\hat{c}(\omega) = E^{\mu}[g(\hat{\tau}_{L,\omega}, X(\hat{\tau}_{L,\omega}, x_0))].$$

Then we have the following.

Theorem 2. *Suppose that* $\gamma \in (0, 1]$. *If*

$$\sum_{k=1}^{n} \mu(|c_{T_k, T_{k+1}, \ldots, T_n}(X(T_k, x_0)) - g(T_k, X(T_k, x_0))| < \varepsilon) = O(\varepsilon^{\gamma}), \ \textit{as}\ \varepsilon \downarrow 0,$$

then for any $\alpha \in (1/2, (1+\gamma)/(2+\gamma))$, *there are* $\Omega_L \in \mathcal{F}$, $L \geqq 1$, *and* $C > 0$ *such that* $P(\Omega_L) \to 1$, $L \to \infty$, *and*

$$|\hat{c}(\omega) - c_{T_0, T_1, \ldots, T_n}| \leqq C L^{-\alpha}\ \textit{for any}\ \omega \in \Omega_L\ \textit{and}\ L \geqq 1.$$

2. The Basic Property of Hörmander Diffusion Processes

Let $J : [0, \infty) \times \mathbf{R}^N \times W_0 \to \mathbf{R}^N \otimes \mathbf{R}^N$, $J(t, x) = (J^i_j(t, x))_{i,j=1,\ldots,N}$ be given by

$$J^i_j(t, x) = \frac{\partial}{\partial x^j} X^i(t, x).$$

Then it has been shown in [6] Sect. 2 that there are $b^{\beta}_{\alpha} : [0, \infty) \times \mathbf{R}^N \times W_0 \to \mathbf{R}$, $\alpha, \beta \in \mathcal{A}^{**}_{\leqq \ell_0}$, such that

$$V_{[\alpha]}(x) = \sum_{\beta \in \mathcal{A}^{**}_{\leqq \ell_0}} b^{\beta}_{\alpha}(t, x) J(t, x)^{-1} V_{[\beta]}(X(t, x)), \qquad \alpha \in \mathcal{A}^{**}_{\leqq \ell_0},$$

and

$$\sup_{x\in\mathbf{R}^N, t\in[0,T]} E^{\mu}[|b_{\alpha}^{\beta}(t,x)|^p] < \infty \qquad \alpha, \beta \in \mathcal{A}^{**}_{\leqq \ell_0}, \ T > 0, \ p \geqq 1.$$

So we see that for any $\xi \in \mathbf{R}^N$,

$$\begin{aligned}
(A(x)\xi,\xi) &= \sum_{\alpha\in\mathcal{A}^{**}_{\leqq\ell_0}} (V_{[\alpha]}(x),\xi)^2 \\
&\leqq \sum_{\alpha\in\mathcal{A}^{**}_{\leqq\ell_0}} \Big(\sum_{\beta\in\mathcal{A}^{**}_{\leqq\ell_0}} b_{\alpha}^{\beta}(t,x)^2\Big)\Big(\sum_{\beta\in\mathcal{A}^{**}_{\leqq\ell_0}} (J(t,x)^{-1}V_{[\beta]}(X(t,x)),\xi)^2\Big) \\
&= \Big(\sum_{\alpha\in\mathcal{A}^{**}_{\leqq\ell_0}} \sum_{\beta\in\mathcal{A}^{**}_{\leqq\ell_0}} b_{\alpha}^{\beta}(t,x)^2\Big)(J(t,x)A(X(t,x))^t J(t,x)\xi,\xi).
\end{aligned}$$

Therefore we see that

$$h(x) \leqq \Big(\sum_{\alpha\in\mathcal{A}^{**}_{\leqq\ell_0}} \sum_{\beta\in\mathcal{A}^{**}_{\leqq\ell_0}} b_{\alpha}^{\beta}(t,x)^2\Big)^N det(J(t,x))^2 h(X(t,x)). \tag{2}$$

Then we have the following.

Proposition 3. (1) $\mu(X(t,x)\in E) = 1$ *for any* $x \in E$ *and* $t > 0$. *In particular,* $p(t,x,y) = 0$, $y \in \mathbf{R}^N \setminus E$, $x \in E$.
(2) *For any* $p > 1$ *and* $T > 0$, *there exists a* $C > 0$ *such that*

$$E^{\mu}[h(X(t,x))^{-p}] \leqq Ch(x)^{-p}, \qquad x \in E, \ t \in [0,T].$$

(3) *For any* $n, m \geqq 0$, $p \in (1,\infty)$, *and* $T > 0$, *there exists a* $C > 0$ *such that*

$$||h(X(t,x))^{-m}||_{W^{n,p}} \leqq Ch(x)^{-(n+m)} \qquad x \in E, \ t \in [0,T].$$

Here $||\cdot||_{W^{n,p}}$ *is the norm of a Sobolev space* $W^{n,p}$ (c.f. Shigekawa [8]).

Proof. The assertions (1) and (2) are easy consequence of Eq. (2). Note that

$$D(h^{-m}(X(t,x))) = -mh^{-(m+1)}(X(t,x)))D(h(X(t,x))).$$

Thus we easily obtain the assertion (3) by induction. ■

By Kusuoka–Stroock [7], we have the following.

Proposition 4. *Let $\delta_0 > 0$ be given by*

$$\delta_0 = (3N(\sup_{x\in\mathbf{R}^N}\sum_{k=1}^{d}|V_k(x)|^2))^{-1}.$$

Then we have the following.

(1) *For any $T > 0$,*

$$\sup_{t\in(0,T],\ x\in\mathbf{R}^N} E^{\mu}[\exp(\frac{2\delta_0}{t}|X(t,x)-x|^2)] < \infty.$$

(2) *For any $T > 0$, $n \geqq 1$, and $p \in (1,\infty)$,*

$$\sup_{t\in(0,T],\ x\in\mathbf{R}^N} t^{n/2}||\exp(\frac{\delta_0}{t}|X(t,x)-x|^2)||_{W^{n,p}} < \infty.$$

Proposition 5. *For any $\gamma \in \mathbf{Z}^N_{\geqq 0}$, there are $g_{\gamma,\alpha_1,\ldots,\alpha_k} \in C_b^\infty(\mathbf{R}^N)$, $k = 1,\ldots,|\gamma|$, $\alpha_i \in \mathcal{A}^{**}_{\leqq \ell_0}$, $i = 1,\ldots,k$, such that*

$$h(x)^{|\gamma|}\frac{\partial^{|\gamma|}}{\partial x^\gamma}f(x) = \sum_{k=1}^{|\gamma|}\sum_{\alpha_1,\ldots,\alpha_k\in\mathcal{A}^{**}_{\leqq \ell_0}} g_{\gamma,\alpha_1,\ldots,\alpha_k}(x)(V_{[\alpha_1]}\cdots V_{[\alpha_k]}f)(x),\quad x\in\mathbf{R}^N$$

for any $f \in C_b^\infty(\mathbf{R}^N)$.

Proof. Let $\tilde{A}(x) = (\tilde{A}_{ij}(x))_{i,j=1,\ldots,N}$ be the cofactor matrix of the matrix $A(x)$ for $x \in \mathbf{R}^N$. Also, let $c_{\alpha,i}(x)$, $x \in \mathbf{R}^N$, $\alpha \in \mathcal{A}^{**}_{\leqq \ell_0}$, $i = 1,\ldots,N$, be given by

$$c_{\alpha,i}(x) = \sum_{j=1}^{N}\tilde{A}_{ij}(x)V^j_{[\alpha]}(x).$$

Then we see that $h, c_{\alpha,i} \in C_b^\infty(\mathbf{R}^N)$, and

$$\sum_{\alpha\in\mathcal{A}^{**}_{\leqq \ell_0}} c_{\alpha,i}(x)(V_{[\alpha]}f)(x) = h(x)\frac{\partial f}{\partial x^i}(x),\qquad i = 1,\ldots,N.$$

So we have the assertion for the case that $|\gamma| = 1$. Since

$$h(x)^{|\gamma|+1}(x)\frac{\partial}{\partial x^i}\frac{\partial^{|\gamma|}}{\partial x^\gamma}f(x)$$
$$= h(x)\frac{\partial}{\partial x^i}(h^{|\gamma|}\frac{\partial^{|\gamma|}}{\partial x^\gamma}f)(x) - |\gamma|\frac{\partial h}{\partial x^i}(x)h^{|\gamma|}(x)\frac{\partial^{|\gamma|}}{\partial x^\gamma}f(x),$$

we have our assertion by induction. ■

Now we have the following lemma.

Lemma 6. *For any $t > 0$, $x \in E$ and $\gamma_0, \gamma_1 \in \mathbf{Z}^N_{\geqq 0}$, there are $k_{\gamma_0,\gamma_1}(t,x) \in W^{\infty,\infty-}$ such that*

$$\int_{\mathbf{R}^N} \partial_x^{\gamma_0}\partial_y^{\gamma_1} p(t,x,y)f(y)dy = E^\mu[h(X(t,x))^{-2(|\gamma_0|+|\gamma_1|)\ell_0}$$
$$\times f(X(t,x))k_{\gamma_0,\gamma_1}(t,x)],$$

$$f \in C_0^\infty(\mathbf{R}^N),$$

and

$$\sup_{t\in(0,T],x\in E} t^{(|\gamma_0|+|\gamma_1|)\ell_0/2}||k_{\gamma_0,\gamma_1}(t,x)||_{W^n,p} < \infty, \quad T > 0,\ n \in \mathbf{N},\ p \in (1,\infty).$$

Here $\partial_x^\gamma = \partial^{|\gamma|}/\partial x^\gamma$ and $\partial_y^\gamma = \partial^{|\gamma|}/\partial y^\gamma$.

Proof. First, by the argument in Shigekawa [8] we see that for $\gamma \in \mathbf{Z}^N_{\geqq 0}$, there are $J_{\gamma,\beta}(t,x) \in W^{\infty,\infty-}$, $t \geqq 0$, $x \in \mathbf{R}^N$, $\beta \in \mathbf{Z}^N_{\geqq 0}$, $|\beta| \leqq |\gamma|$, such that

$$\partial_x^\gamma(f(X(t,x)) = \sum_{\beta\in\mathbf{Z}^N_{\geqq 0},\ |\beta|\leqq|\gamma|} (\partial_x^\beta f)(X(t,x))J_{\gamma,\beta}(t,x),$$

and

$$\sup_{t\in(0,T],x\in\mathbf{R}^N} ||J_{\gamma,\beta}(t,x)||_{W^n,p} < \infty, \qquad T > 0,\ n \in \mathbf{N},\ p \in (1,\infty).$$

Then we have for any $x \in E$ and $f \in C_0^\infty(\mathbf{R}^N)$,

$$\int_{\mathbf{R}^N} \partial_x^{\gamma_0}\partial_y^{\gamma_1} p(t,x,y)f(y)dy$$
$$= (-1)^{|\gamma_1|}\int_{\mathbf{R}^N} \partial_x^{\gamma_0} p(t,x,y)(\partial_y^{\gamma_1} f)(y)dy$$
$$= (-1)^{|\gamma_1|}\partial_x^{\gamma_0} E^\mu[(\partial_y^{\gamma_1} f)(X(t,x))]$$

$$= (-1)^{|\gamma_1|} \sum_{\beta \in \mathbf{Z}^N_{\geqq 0},\ |\beta| \leqq |\gamma_0|} E^{\mu}[(\partial_x^{\gamma_1+\beta} f)(X(t,x)) J_{\gamma_0,\beta}(t,x)]$$

$$= (-1)^{|\gamma_1|} \sum_{\beta \in \mathbf{Z}^N_{\geqq 0},\ |\beta| \leqq |\gamma_0|} \sum_{k=0}^{|\gamma_1|+|\beta|} \sum_{\alpha_1,\dots,\alpha_k \in \mathcal{A}^{**}_{\leqq \ell_0}} E^{\mu}[h(X(t,x))^{-(|\gamma_1|+|\beta|)}$$
$$\times g_{\gamma_1+\beta,\alpha_1,\dots,\alpha_k}(X(t,x)) J_{\gamma_0,\beta}(t,x)(V_{[\alpha_1]} \cdots V_{[\alpha_k]} f)(X(t,x))].$$

So by the integration parts formula in [6] Lemma 8 and by Propositions 3 and 5, we have our assertion. ■

Proposition 7. *For any $t > 0$, $x \in E$ and $\gamma_0, \gamma_1 \in \mathbf{Z}^N_{\geqq 0}$,*

$$\partial_x^{\gamma_0} \partial_y^{\gamma_1} p(t,x,y) = 0 \ a.e. y \in \mathbf{R}^N \setminus E.$$

Moreover, for any $\gamma_0, \gamma_1 \in \mathbf{Z}^N_{\geqq 0}$, $p \in (1,\infty)$, $T > 0$, and $m \in \mathbf{Z}$ with $m \leqq 2(|\gamma_0| + |\gamma_1|)$,

$$\sup\{t^{(|\gamma_0|+\gamma_1|)\ell_0/2} h(x)^{2(|\gamma_0|+|\gamma_1|)\ell_0 - m} (\int_E h(y)^{pm} \exp(\frac{p\delta_0}{t}|y-x|^2)$$
$$\times \frac{|\partial_x^{\gamma_0}\partial_y^{\gamma_1} p(t,x,y)|^p}{p(t,x,y)^{p-1}} dy)^{1/p};$$
$$t \in (0,T],\ x \in E\} < \infty.$$

Proof. Let

$$\varphi_{t,x}(y) = \exp(\frac{\delta_0}{t}|y-x|^2), \qquad x, y \in \mathbf{R}^N,\ t > 0.$$

Then we have for any $\varepsilon > 0$, $f \in C_0^\infty(\mathbf{R}^N)$ and $x \in E$

$$\int_{\mathbf{R}^N} \partial_x^{\gamma_0}\partial_y^{\gamma_1} p(t,x,y) f(y)(\varepsilon + h(y))^m \varphi_{t,x}(y) dy$$
$$= E^{\mu}[h(X(t,x))^{-2(|\gamma_0|+|\gamma_1|)\ell_0} f(X(t,x))(\varepsilon + h(X(t,x)))^m$$
$$\times \varphi_{t,x}(X(t,x))) k_{\gamma_0,\gamma_1}(t,x)].$$

By Propositions 3 and 4, we see that

$$\int_{\mathbf{R}^N} \partial_x^{\gamma_0}\partial_y^{\gamma_1} p(t,x,y) f(y) h(y)^m \varphi_{t,x}(y) dy$$
$$= E^{\mu}[h(X(t,x))^{m-2(|\gamma_0|+|\gamma_1|)\ell_0} f(X(t,x)) \varphi_{t,x}(X(t,x))) k_{\gamma_0,\gamma_1}(t,x)].$$

Let $k'(t,x) = h(X(t,x))^{m-2(|\gamma_0|+|\gamma_1|)\ell_0}\varphi_{t,x}(X(t,x))k_{\gamma_0,\gamma_1}(t,x)$. Then we see that

$$\sup_{t\in(0,T],x\in E} t^{(|\gamma_0|+|\gamma_1|)\ell_0/2}h(x)^{2(|\gamma_0|+|\gamma_1|)\ell_0-m}E^{\mu}[|k'(t,x)|^p]^{1/p}$$
$$< \infty,\ T > 0,\ p \in (1,\infty).$$

Note that there is a Borel function $\tilde{k}(t,x) : \mathbf{R}^N \to \mathbf{R}$, $t \in (0,T]$, $x \in E$, such that

$$E^{\mu}[k'(t,x)|\sigma\{X(t,x))\}] = \tilde{k}(t,x)(X(t,x)), \qquad t \in (0,T],\ x \in E.$$

Then we have

$$\begin{aligned}
&\int_{\mathbf{R}^N} \partial_x^{\gamma_0}\partial_y^{\gamma_1} p(t,x,y)f(y)h(y)^m\varphi_{t,x}(y)dy \\
&= E^{\mu}[k'(t,x)f(X(t,x))] = E^{\mu}[\tilde{k}(t,x)(X(t,x))f(X(t,x))] \\
&= \int_{\mathbf{R}^N} f(y)\tilde{k}(t,x)(y)p(t,x,y)dy,
\end{aligned}$$

for any $f \in C_0^\infty(\mathbf{R}^N)$. This implies that $\partial_x^{\gamma_0}\partial_y^{\gamma_1} p(t,x,y)h(y)^m\varphi_{t,x}(y) = \tilde{k}(t,x)(y)p(t,x,y)$ *a.e.y*, $t \geqq 0$, $x \in E$. Therefore letting $m = 0$, we have the first assertion. Since

$$\begin{aligned}
&\int_E h(y)^{pm}\varphi_{t,x}(y)^p \frac{|\partial_y^{\gamma} p(t,x,y)|^p}{p(t,x,y)^{p-1}}dy = \int_E |\tilde{k}(t,x,y)|^p p(t,x,y)dy \\
&= E^{\mu}[|\tilde{k}(t,x)(X(t,x))|^p] \leqq E^{\mu}[|k'(t,x)|^p],
\end{aligned}$$

we have our assertion. ■

Proposition 8. *For any* $T > 0$, *there is a* $C > 0$ *such that*

$$\begin{aligned}
p(t,x,y) &\leqq Ct^{-(N+1)\ell_0/2}h(x)^{-2(N+1)\ell_0} \\
&\quad \times \exp(-\frac{2\delta_0}{t}|y-x|^2),\quad t \in (0,T],\ x,y \in E
\end{aligned}$$

and

$$\begin{aligned}
p(t,x,y) &\leqq Ct^{-(N+1)\ell_0/2}h(y)^{-2(N+1)\ell_0} \\
&\quad \times \exp(-\frac{2\delta_0}{t}|y-x|^2),\quad t \in (0,T],\ x,y \in E.
\end{aligned}$$

In particular, for any $T > 0$ *and* $m \geqq 1$, *there is a* $C > 0$ *such that*

$$\begin{aligned}
p(t,x,y) &\leqq Ct^{-(N+1)\ell_0/2}h(x)^{-2(N+1)\ell_0}(1+|x|^2)^m(1+|y|^2)^{-m}, \\
&\quad t \in (0,T],\ x,y \in E.
\end{aligned}$$

Proof. Let C_0

$$= \sup\{t^{\ell_0/2}h(x)^2(\int_E \exp(\frac{2(N+1)\delta_0}{t}|y-x|^2)\frac{|\partial_{y^i}p(t,x,y)|^{N+1}}{p(t,x,y)^N}dy)^{1/(N+1)};$$

$$t \in (0,T],\ x \in E, \varepsilon > 0\}.$$

Let

$$\rho_\varepsilon(t,x,y) = (p(t,x,y) + \varepsilon \exp(-(1+\frac{2\delta_0}{t})|y-x|^2))^{1/(N+1)}.$$

Then we see that

$$(\int_{\mathbf{R}^N} \exp(\frac{2\delta_0}{t}|y-x|^2)|\frac{\partial}{\partial y^i}\rho_\varepsilon(t,x,y)|^{N+1}dy)^{1/(N+1)}$$

$$= (N+1)^{-1}(\int_{\mathbf{R}^N} \exp(\frac{2\delta_0}{t}|y-x|^2)$$

$$\times \frac{|\partial_{y^i}(p(t,x,y) + \varepsilon \exp(-(1+\frac{2\delta_0}{t})|y-x|^2))|^{N+1}}{(p(t,x,y) + \varepsilon \exp(-(1+\frac{2\delta_0}{t})|y-x|^2))^N}dy)^{1/(N+1)}$$

$$\leqq (\int_{\mathbf{R}^N} \exp(\frac{2\delta_0}{t}|y-x|^2)$$

$$\times \frac{|\partial_{y^i}p(t,x,y)|^{N+1}}{(p(t,x,y) + \varepsilon \exp(-(1+\frac{2\delta_0}{t})|y-x|^2))^N}dy)^{1/(N+1)}$$

$$+ (\int_{\mathbf{R}^N} \exp(\frac{2\delta_0}{t}|y-x|^2)$$

$$\times \frac{|\partial_{y^i}(\varepsilon \exp(-(1+\frac{2\delta_0}{t})|y-x|^2))|^{N+1}}{(p(t,x,y) + \varepsilon \exp(-(1+\frac{2\delta_0}{t})|y-x|^2))^N}dy)^{1/(N+1)}$$

$$\leqq C_0 t^{-\ell_0/2}h(x)^{-2} + (\varepsilon \int_{\mathbf{R}^N} (2|y^i - x^i|)^{N+1}(1+\frac{1}{t}])^{N+1}$$

$$\times \exp(-|y-x|^2))dy)^{1/(N+1)}.$$

Also, we have

$$(\int_{\mathbf{R}^N} \exp(\frac{2\delta_0}{t}|y-x|^2)\rho_\varepsilon(t,x,y)^{N+1}dy)^{1/(N+1)}$$

$$= (\int_{\mathbf{R}^N} \exp(\frac{2\delta_0}{t}|y-x|^2)(p(t,x,y)$$

$$+ \varepsilon \exp(-(1+\frac{2\delta_0}{t})|y-x|^2))dy)^{1/(N+1)}$$

$$= (E^\mu[\exp(\frac{2\delta_0}{t}|X(t,x)-x|^2)] + \pi^N\varepsilon)^{1/(N+1)},$$

and

$$(\int_{\mathbf{R}^N}(|\partial_{y_i}(\exp(\frac{2\delta_0}{(N+1)t}|y-x|^2))|\rho_\varepsilon(t,x,y))^{N+1}dy)^{1/(N+1)}$$
$$=(\int_{\mathbf{R}^N}(\frac{4\delta_0|y_i-x_i|}{t})^{N+1}\exp(\frac{2\delta_0}{(N+1)t}|y-x|^2)(p(t,x,y)$$
$$+\varepsilon\exp(-(1+\frac{2\delta_0}{t})|y-x|^2))dy)^{1/(N+1)}$$
$$\leqq\frac{4\delta_0}{t}E^\mu[|X(t,x)-x|^{N+1}\exp(\frac{2\delta_0}{t}|X(t,x)-x|^2)]^{1/(N+1)}$$
$$+\varepsilon\frac{4\delta_0}{t}(\int_{\mathbf{R}^N}|y_i|^{N+1}\exp(-|y|^2))dy)^{1/(N+1)}.$$

Then by Sobolev's inequality, we see that there is a constant $C>0$ such that

$$\sup_{y\in\mathbf{R}^N}(\exp(\frac{2\delta_0}{t}|y-x|^2)((p(t,x,y)+\varepsilon\exp(-|y|^2)))^{1/(N+1)}$$
$$\leqq C(C_0t^{-\ell_0/2}h(x)^{-2\ell_0}+Ct^{-1/2}+C\varepsilon(1+\frac{1}{t})).$$

So letting $\varepsilon\downarrow 0$, we have our first assertion.

Let

$$\tilde{\rho}_\varepsilon(t,x,y)=(p(t,x,y)h(y)^{2(N+1)\ell_0}+\varepsilon\exp(-(1+\frac{2\delta_0}{t})|y-x|^2))^{1/(N+1)}.$$

Then similarly we can show that

$$\int_{\mathbf{R}^N}(\exp(\frac{2\delta_0}{(N+1)t}|y-x|^2)\tilde{\rho}_\varepsilon(t,x,y))^{N+1}$$
$$+\sum_{i=1}^N|\partial_{y_i}(\exp(\frac{2\delta_0}{(N+1)t}|y-x|^2)\tilde{\rho}_\varepsilon(t,x,y))|^{N+1})dy\leqq Ct^{-\ell_0/2},$$
$$t\in(0,T],\ x\in E.$$

So we have our second assertion.

Finally note that

$$|\log(1+|x|^2)-\log(1+|y|^2)|\leqq|\int_{|y|}^{|x|}\frac{2t}{1+t^2}dt|\leqq|x-y|\leqq\frac{1}{\varepsilon}+\varepsilon|x-y|^2,$$
$$x,y\in\mathbf{R}^N,\ \varepsilon>0.$$

So we have the final assertion. ■

Proposition 9. *Let* $\delta \in (0, 1/N)$, $\alpha, \beta \in \mathbf{Z}^N_{\geqq 0}$ *and* $T > 0$. *Then there are* $C > 0$ *and* $q > 0$ *such that*

$$|\partial_x^\alpha \partial_y^\beta p(t,x,y)| \leqq Ct^{-(|\alpha|+|\beta|+1)\ell_0/2} h(x)^{-2(|\alpha|+|\beta|+1)\ell_0} p(t,x,y)^{1-\delta},$$

$$x, y \in E,\ t \in (0, T],$$

and

$$|\partial_x^\alpha \partial_y^\beta p(t,x,y)| \leqq Ct^{-(|\alpha|+|\beta|+1)\ell_0/2} h(y)^{-2(|\alpha|+|\beta|+1)\ell_0} p(t,x,y)^{1-\delta},$$

$$x, y \in E,\ t \in (0, T].$$

Proof. Let $p = 1/\delta > N$, and let

$$\rho_\varepsilon(t,x,y) = \frac{\partial_x^\alpha \partial_y^\beta p(t,x,y)}{(p(t,x,y)+\varepsilon)^{1-\delta}}$$

for $\varepsilon > 0$. Then we see by Proposition 7 that there is a $C_1 > 0$ such that

$$\left(\int_{\mathbf{R}^N} |\rho_\varepsilon(t,x,y)|^p dy\right)^{1/p} = \left(\int_{\mathbf{R}^N} \frac{|\partial_x^\alpha \partial_y^\beta p(t,x,y)|^p}{(p(t,x,y)+\varepsilon)^{p-1}} dy\right)^{1/p}$$
$$\leqq C_1 t^{-(|\alpha|+|\beta|)\ell_0/2} h(y)^{-2(|\alpha|+|\beta|)\ell_0}, \qquad \varepsilon > 0,\ t \in (0, T],\ x \in E.$$

Also, we have

$$\left(\int_{\mathbf{R}^N} |\partial_{y^i} \rho_\varepsilon(t,x,y)|^p dy\right)^{1/p}$$
$$\leqq \left(\int_{\mathbf{R}^N} \frac{|\partial_x^\alpha \partial_y^\beta \partial_{y^i} p(t,x,y)|^p}{(p(t,x,y)+\varepsilon)^{p-1}} dy\right)^{1/p}$$
$$+(1-\delta)\left(\int_{\mathbf{R}^N} \frac{|\partial_x^\alpha \partial_y^\beta p(t,x,y)|^{2p}}{(p(t,x,y)+\varepsilon)^{2p-1}} dy\right)^{1/(2p)} \left(\int_{\mathbf{R}^N} \frac{|\partial_{y_i} p(t,x,y)|^{2p}}{(p(t,x,y)+\varepsilon)^{2p-1}} dy\right)^{1/(2p)}.$$

So we see by Proposition 7 that there is a $C_2 > 0$ such that

$$\left(\int_{\mathbf{R}^N} |\partial_{y_i} \rho_\varepsilon(t,x,y)|^p dy\right)^{1/p}$$
$$\leqq C_2 t^{-(|\alpha|+|\beta|+1)\ell_0/2} h(y)^{-2(|\alpha|+|\beta|+1)\ell_0}, \quad \varepsilon > 0,\ t \in (0, T],\ x \in E.$$

So by Sobolev's inequality, we see that there is a $C_3 > 0$ such that

$$\sup_{y \in \mathbf{R}^N} |\rho_\varepsilon(t, x, y)| \leqq C_3 t^{-(|\alpha|+|\beta|+1)\ell_0/2} h(x)^{-2(|\alpha|+|\beta|+1)},$$

$$\varepsilon > 0, \ t \in (0, T], \ x \in E.$$

Letting $\varepsilon \downarrow 0$, we have the first assertion.

Let

$$\tilde{\rho}_\varepsilon(t, x, y) = \frac{\partial_x^\alpha \partial_y^\beta p(t, x, y)}{(p(t, x, y) + \varepsilon)^{1-\delta}} h(y)^{2(|\alpha|+|\beta|+1)}$$

for $\varepsilon > 0$. Then a similar argument implies that there is a $C_4 > 0$ such that

$$\sup_{y \in \mathbf{R}^N} |\tilde{\rho}_\varepsilon(t, x, y)| \leqq C_4 t^{-(|\alpha|+|\beta|+1)\ell_0/2},$$

$$\varepsilon > 0, \ t \in (0, T], \ x \in E.$$

So we have the second assertion. ■

Proposition 10. *Let $m \geqq 0$, $\alpha, \beta \in \mathbf{Z}^N_{\geqq 0}$, $p \in [1, \infty)$, $\delta \in (0, 1)$ and $T > 0$. Then there is a $C > 0$ such that*

$$\int_{\mathbf{R}^N} |\partial_t^m \partial_x^\alpha \partial_y^\beta p(t - s, x, y)|^p p(s, x_0, x) dx$$
$$\leqq C(t - s)^{p(|\alpha|+|\beta|+2m+2)\ell_0/2} p(t, x_0, y)^{1-\delta}$$

for any $t \in (0, T]$, $s \in [0, t)$, $y \in \mathbf{R}^N$.

Proof. First note that

$$\partial_t p(t, x, y) = L_x p(t, x, y), \qquad \text{where } L = \frac{1}{2}\sum_{k=1}^d V_k^2 + V_0.$$

So it is sufficient to prove the case $m = 0$.

Let $r = 1/(1 - \delta)$. Since $p > 1 - \delta$, we see by Propositions 8 and 9, that there are $C > 0$ and $b > 0$ such that

$$|\partial_x^\alpha \partial_y^\beta p(t - s, x, y)|^p \leqq C(t - s)^{p(|\alpha|+|\beta|+2)\ell_0/2} h(x)^{-b} p(t - s, x, y)^{1-\delta},$$

for any $t \in (0, T]$, $s \in [0, t)$, $x \in E$, $y \in \mathbf{R}^N$. So we see that

$$\int_{\mathbf{R}^N} |\partial_x^\alpha \partial_y^\beta p(t-s,x,y)|^p p(s,x_0,x)dx$$
$$\leqq C(t-s)^{p(|\alpha|+|\beta|+2)\ell_0/2} \int_{\mathbf{R}^N} h(z)^{-b} p(t-s,z,y)^{1/r} p(s,x_0,z)dz$$
$$\leqq C(t-s)^{p(|\alpha|+|\beta|+2)\ell_0/2} (\int_{\mathbf{R}^N} (h(z)^{-b/\delta} p(s,x_0,z)dz)^\delta$$
$$\times (\int_{\mathbf{R}^N} p(t-s,z,y)p(s,x_0,z)dz)^{1-\delta}.$$

Since

$$\int_{\mathbf{R}^N} p(t-s,z,y)p(s,x_0,z)dz = p(t,x_0,y),$$

we have our assertion. ■

Proposition 11. *Let $a \in (0,1]$, and $b \in (0,a)$. Then we have*

$$\int_{\mathbf{R}^N} p(s,x_0,x)^a p(t-s,x,y)^b \phi(x)dx$$
$$\leqq p(t,x_0,y)^b (\int_E dx p(s,x_0,x)^{(a-b)/(1-b)} \phi(x)^{1/(1-b)})^{1-b}$$

for any $t > s \geqq 0$, and non-negative measurable function $\phi : E \to [0,\infty)$.

Proof. Let $\delta = (a-b)/(1-b)$, $p = 1/b$, and $q = 1/(1-b)$. Then we see that $1-\delta = (1-a)/(1-b)$ and $a-\delta = b(1-a)/(1-b)$, and so we have

$$\int_{\mathbf{R}^N} p(s,x_0,x)^a p(t-s,x,y)^b \phi(x)dx$$
$$= \int_{\mathbf{R}^N} p(s,x_0,x)^\delta p(s,x_0,x)^{(1-\delta)/p} p(t-s,x,y)^{1/p} \phi(x)dx$$
$$\leqq (\int_E p(s,x_0,x)^\delta p(s,x_0,x)^{1-\delta} p(t-s,x,y)dx)^{1/p}$$
$$\times (\int_E p(s,x_0,x)^\delta \phi(x)^q dx)^{1/q}$$
$$= p(s,x_0,y)^b (\int_E p(s,x_0,x)^{(a-b)/(1-b)} \phi(x)^{1/(1-b)} dx)^{1-b}.$$

This proves our assertion. ■

Proposition 12. *Let $p \geqq 1$, $m \geqq 1$. $\alpha, \beta \in \mathbf{Z}^N_{\geqq 0}$, $T > 0$, $a \in (0, 1/p]$ and $b \in (a - 1/N, a)$. Then there is a $C > 0$ such that*

$$\int_{\mathbf{R}^N} |\partial_x^\alpha(p(s,x_0,x)^a)|^p |\partial_x^\beta p(t-s,x,y)|^p dx$$
$$\leqq Cs^{-p(|\alpha|+1)\ell_0/2}(t-s)^{-p(|\beta|+2)\ell_0/2} p(t,x_0,y)^{pb}(1+|y|^2)^{-m}$$

for any $y \in E$ and $s,t \in (0,T]$ with $s<t$.

Proof. Let $\delta = (a-b)/2 < 1/N$. Note that $\partial_x^\alpha(p(s,x_0,x)^a)$ is a linear combination of $a(a-1)\cdots(a-m+1)p(s,x_0,x)^{a-m}\partial_x^{\alpha_1} p(s,x_0,x)\cdots\partial_x^{\alpha_m} p(s,x_0,x)$, $m=1,\ldots,|\alpha|$, $\alpha_k \in \mathbf{Z}_{\geq 0}$, $|\alpha_k| \geqq 1$, $k=1,\ldots,m$, $\alpha_1+\cdots+\alpha_m=\alpha$.

Then by Propositions 9, we see that there is a $C_1>0$ such that

$$|\partial_x^\alpha(p(s,x_0,x)^a)||\partial_x^\beta p(t-s,x,y)|$$
$$\leqq C_1 s^{-(|\alpha|+1)\ell_0/2}(t-s)^{-(|\beta|+1)\ell_0/2} h(x)^{-2(|\beta|+1)\ell_0} p(s,x_0,x)^{a-\delta}$$
$$\times\, p(t-s,x,y)^{1-\delta}$$

for any $a \in (0,1/p]$, $b \in (a-1/N,a)$, $x,y \in E$ and $s,t \in [0,T]$ with $s<t$. By Propositions 8, we see that there is a $C_2>0$ such that

$$p(t-s,x,y)^{1-\delta-b} \leqq C_2(t-s)^{-\ell_0/2} h(x)^{-2(N+1)\ell_0}$$
$$\times\,(1+|x|^2)^m(1+|y|^2)^{-(1-\delta-b)m}$$

for any $x,y \in E$ and $s,t \in [0,T]$ with $s<t$. So we have

$$|\partial_x^\alpha(p(s,x_0,x)^a)||\partial_x^\beta p(t-s,x,y)|$$
$$\leqq C_1C_2 s^{-(|\alpha|+1)\ell_0/2}(t-s)^{-(|\beta|+2)\ell_0/2} h(x)^{-2(|\beta|+N+2)\ell_0} p(s,x_0,x)^{a-\delta}$$
$$\times\, p(t-s,x,y)^b(1+|x|^2)^m(1+|y|^2)^{-(1-(a+b)/2)m}$$

Note that $pb < p(a-\delta) < 1$, and so we have

$$\int_E (h(x)^{-2(|\beta|+N+2)\ell_0} p(s,x_0,x)^{a-\delta} p(t-s,x,y)^b (1+|x|^2)^m)^p dx$$
$$= \int_E p(s,x_0,x)^{p(a-\delta-b)/(1-pb)} p(s,x_0,y)^{pb(1-p(a-\delta))/(1-pb)}$$
$$\times\, p(t-s,x,y)^{pb}(h(x)^{-2p(|\beta|+N+2)\ell_0}(1+|x|^2)^{mp}\, dx$$
$$\leqq \Big(\int_E p(s,x_0,x)^{p(a-\delta-b)/(1-pb)} p(s,x_0,y)^{(1-p(a-\delta))/(1-pb)}$$
$$\times\, p(t-s,x,y)\, dx\Big)^{pb}\Big(\int_E p(s,x_0,x)^{p(a-\delta-b)/(1-pb)}$$
$$\times\, h(x)^{-p(|\beta|+N+2)/(1-pb)}(1+|x|^2)^{mp/(1-pb)}\, dx\Big)^{1-pb}$$

$$= p(t, x_0, y)^{pb}\Big(\int_E (1 + |x|^2)^{-N} p(s, x_0, x)^{p(a-\delta-b)/(1-pb)}$$
$$\times h(x)^{-2p(|\beta|+N+2)/(1-pb)}(1 + |x|^2)^{mp/(1-pb)+N}\, dx\Big)^{1-pb}$$
$$\leqq p(t, x_0, y)^{pb}\Big(\int_E (1 + |x|^2)^{-N} p(s, x_0, x)h(x)^{-p(|\beta|+N+2)/\ell_0(p(a-\delta-b))}$$
$$\times (1 + |x|^2)^{(mp+N(1-pb))/(p(a-\delta-b))}\, dx\Big)^{p(a-\delta-b)}$$
$$\times \Big(\int_E (1 + |x|^2)^{-N}\, dx\Big)^{(1-p(a-\delta-b))/(1-pb)}$$
$$= p(t, x_0, y)^{pb}\Big(\int_E (1 + |x|^2)^{-N}\, dx\Big)^{(1-p\delta)/(1-pb)}$$
$$\times E^{\mu}[h(X(s, x_0))^{-(|\beta|+N+2)\ell_0/\delta}(1 + |X(s, x_0)|^2)^{(mp+N(1-pb))/(p\delta)}]^{p\delta}.$$

So by Proposition 3, we have our assertion. ■

3. Stochastic Mesh and Random Norms

Let $\mathcal{F}_t^{(L)}$, $t \geqq 0$, $L = 0, 1, \ldots, \infty$ be sub σ-algebra of $\mathcal{F}$ given by

$$\mathcal{F}_t^{(L)} = \sigma\{X_\ell(s);\ s \in [0, t],\ \ell = 1, 2, \ldots .L\},$$

and

$$\mathcal{F}_t^{(\infty)} = \sigma\{X_\ell(s);\ s \in [0, t],\ \ell = 1, 2, \ldots\}.$$

Let ν_t, $t \geqq 0$, be the probability law of $X(t, x_0)$ under μ. Then we see that ν_0 is the probability measure concentrated in x_0, and $\nu_t(dx) = p(t, x_0, x)dx$, $t > 0$.

Then for any $t > s \geqq 0$, we can define a linear contraction map $P_{s,t} : L^1(E; d\nu_t) \to L^1(E; d\nu_s)$ by

$$(P_{s,t} f)(x) = \int_E p(t-s, x, y) f(y)dy, \qquad x \in E,\ f \in L^1(E; d\nu_t).$$

Proposition 13. *Let $t > s \geqq 0$, $\alpha \in \mathbf{Z}_{\geqq 0}^N$ and bounded measurable function $f : E \to \mathbf{R}$. Then we have*

$$E[\partial_x^\alpha (Q_{s,t}^{(L)} f)(x)|\mathcal{F}_s^{(\infty)}] = \partial_x^\alpha (P_{s,t} f)(x), \qquad \nu_s - a.e.x.$$

and

$$E[|\partial_x^\alpha (Q_{s,t}^{(L)} f)(x) - \partial_x^\alpha (P_{s,t} f)(x)|^2 | \mathcal{F}_s^{(\infty)}] \leqq \frac{1}{L} \int_E \frac{(\partial_x^\alpha p(t-s, x, y))^2 |f(y)|^2}{q_{s,t}^{(L)}(y)}\, dy.$$

Proof. Note that

$$E[\partial_x^\alpha (Q_{s,t}^{(L)} f)(x)|\mathcal{F}_s^{(\infty)}] = \frac{1}{L}\sum_{\ell=1}^{L}\int_E \frac{\partial_x^\alpha p(t-s,x,y)f(y)}{q_{s,t}^{(L)}(y)} p(t-s,X_\ell(s),y)\,dy$$

$$= \int_E \partial_x^\alpha p(t-s,x,y)f(y)\,dy = \partial_x^\alpha (P_{s,t}f)(x).$$

This implies the first assertion.

Let

$$m_\ell = \frac{1}{L}\int_E \frac{\partial_x^\alpha p(t-s,x,y)f(y)}{q_{s,t}^{(L)}(y)} p(t-s,X_\ell(s),y)\,dy$$

and

$$d_\ell = \frac{1}{L}\frac{\partial_x^\alpha p(t-s,x,X_\ell(t))f(X_\ell(t))}{q_{s,t}^{(L)}(X_\ell(t))} - m_\ell$$

for $\ell = 1,\ldots,L$. Then we see that

$$E[d_\ell|\mathcal{F}_s^{(\infty)} \vee \mathcal{F}_t^{(\ell-1)}] = 0, \qquad \ell = 1,\ldots,L.$$

Here we let $\mathcal{F}_t^{(0)} = \{\emptyset,\Omega\}$. Moreover, we have

$$\sum_{\ell=1}^{L} d_\ell = \partial_x^\alpha (Q_{s,t}^{(L)} f)(x) - \partial_x^\alpha (P_{s,t}f)(x)$$

So we see that

$$E[|\partial_x^\alpha (Q_{s,t}^{(L)} f)(x) - \partial_x^\alpha (P_{s,t}f)(x)|^2|\mathcal{F}_s^{(\infty)}] \leqq E[(\sum_{\ell=1}^{L}|d_\ell|^2)|\mathcal{F}_s^{(\infty)}]$$

$$\leqq \sum_{\ell=1}^{L} E[(\frac{1}{L}\frac{\partial_x^\alpha p(t-s,x,X_\ell(t))f(X_\ell(t))}{q_{s,t}^{(L)}(X_\ell(t))})^2|\mathcal{F}_s^{(\infty)}]$$

$$\leqq \frac{1}{L^2}\sum_{\ell=1}^{L}\int_E \frac{(\partial_x^\alpha p(t-s,x,y))^2|f(y)|^2}{q_{s,t}^{(L)}(y)^2} p(t-s,X_\ell(s),y)\,dy$$

$$= \frac{1}{L}\int_E \frac{(\partial_x^\alpha p(t-s,x,y))^2|f(y)|^2}{q_{s,t}^{(L)}(y)}\,dy.$$

So we have the second assertion. ■

Now let $M_t^{(L)} : m(E) \times \Omega \to \mathbf{R}$, and $N_t^{(L)} : m(E) \times \Omega \to [0, \infty)$, $t \geqq 0$, $L \geqq 1$, be random functionals given by

$$M_t^{(L)}(f) = M_t^{(L)}(f; \omega) = \frac{1}{L} \sum_{\ell=1}^{L} f(X_\ell(t)), \qquad f \in m(E),$$

and

$$N_t^{(L)}(f) = N_t^{(L)}(f; \omega) = M_t^{(L)}(|f|) = \frac{1}{L} \sum_{\ell=1}^{L} |f(X_\ell(t))|, \qquad f \in m(E).$$

Then we see that $M_t^{(L)}$ is a linear function and $N_t^{(L)}$ is a semi-norm in $m(E)$.

Proposition 14. *Let $t > s \geqq 0$ and $L \geqq 1$ (1) For any $f \in m(E)$,*

$$M_s^{(L)}(Q_{s,t}^{(L)} f) = M_t(f).$$

(2) *For any $f \in m(E)$*

$$N_s^{(L)}(Q_{s,t}^{(L)} f) \leqq N_t(f).$$

Proof. Suppose that $f \in m(E)$. Then we have

$$M_s^{(L)}(Q_{s,t}^{(L)} f) = \frac{1}{L} \sum_{\ell=1}^{L} \frac{1}{L} \sum_{k=1}^{L} \frac{p(t-s, X_\ell(s), X_k(t)) f(X_k(t))}{q_{s,t}^{(L)}(X_k(t))}$$

$$= \frac{1}{L} \sum_{k=1}^{L} (\frac{1}{L} \sum_{\ell=1}^{L} \frac{p(t-s, X_\ell(s), X_k(t)) f(X_k(t))}{q_{s,t}^{(L)}(X_k(t))}) = M_t(f).$$

So we have the assertion (1).

The second assertion is an easy consequence of the assertion (1). ■

Proposition 15. (1) *Let $T > 0$ and $m \geqq 1$. Then there is a $C > 0$ such that*

$$\frac{1}{L} \sum_{\ell=1}^{L} E[((Q_{s,t}^{(L)} f)(X_\ell(s)) - (P_{s,t} f)(X_\ell(s))^2 | \mathcal{F}_s^{(\infty)}]$$
$$\leqq \frac{C}{L}(t-s)^{-(N+1)\ell_0/2} \max_{\ell=1\ldots,L} h(X_\ell(s))^{-2(N+1)}(1 + |X_\ell(s)|^2)^m)$$
$$\times \int_E f(y)^2 (1 + |y|^2)^{-m}\, dy \quad a.s.$$

for any $L \geqq 1$ *and* $s, t \in [0, T]$ *with* $s < t$.
In particular,

$$E[N_s^{(L)}(Q_{s,t}^{(L)} f - P_{s,t} f)^2]$$
$$\leqq \frac{C}{L}(t-s)^{-(N+1)\ell_0/2} E[\max_{\ell=1\ldots,L} h(X_\ell(s))^{-2(N+1)}(1+|X_\ell(s)|^2)^m]$$
$$\times \int_E f(y)^2(1+|y|^2)^{-m} dy$$

for any $L \geqq 1$ *and* $s, t \in [0, T]$ *with* $s < t$.
(2) *For any* $\varepsilon > 0$ *and* $T > 0$,

$$\overline{\lim_{L\to\infty}} L^{-\varepsilon} \sup_{s\in[0,T]} E[\max_{\ell=1\ldots,L} h(X_\ell(s))^{-2(N+1)}(1+|X_\ell(s)|^2)^m] = 0$$

Proof. By Proposition 13, we see that

$$\frac{1}{L}\sum_{\ell=1}^{L} E[((Q_{s,t}^{(L)} f)(X_\ell(s)) - (P_{s,t} f)(X_\ell(s))^2|\mathcal{F}_s^{(\infty)}]$$
$$\leqq \frac{1}{L^2}\sum_{\ell=1}^{L}\int_E \frac{p(t-s, X_\ell(s), y)^2 f(y)^2}{q_{s,t}^{(L)}(y)} dy$$
$$\leqq \frac{1}{L^2}\sum_{\ell=1}^{L}\int_E (\max_{\ell'=1,\ldots,L} p(t-s, X_{\ell'}(s), y)) \frac{p(t-s, X_\ell(s), y) f(y)^2}{q_{s,t}^{(L)}(y)} dy$$
$$= \frac{1}{L}\int_E (\max_{\ell=1,\ldots,L} p(t-s, X_\ell(s), y)) f(y)^2 dy.$$

Then by Proposition 8 we have the assertion (1).
Let $\varepsilon > 0$. Let us take $p > 1/\varepsilon$. Then we have

$$E[\max_{\ell=1\ldots,L} h(X_\ell(s))^{-2(N+1)}(1+|X_\ell(s)|^2)^m]$$
$$\leqq E[(\sum_{\ell=1}^{L}(h(X_\ell(s))^{-2(N+1)}(1+|X_\ell(s)|^2)^m)^p)^{1/p}]$$
$$\leqq E[(\sum_{\ell=1}^{L}(h(X_\ell(s))^{-2(N+1)}(1+|X_\ell(s)|^2)^m))^p]^{1/p}$$
$$= L^{1/p} E^\mu[h(X(s,x_0))^{-2p(N+1)}(1+|X(s,x_0)|^2)^{mp}]^{1/p}$$
$$\leqq L^{1/p} E^\mu[h(X(s,x_0))^{-4p(N+1)}]^{1/(2p)} E^\mu[(1+|X(s,x_0)|^2)^{2pm}]^{1/(2p)}$$

So we have the assertion (2) by Proposition 3.

4. Application 1

Let $r \geqq 0$, and let $\mathcal{B}_r$ be the set of Borel measurable functions $f : \mathbf{R}^N \to \mathbf{R}$ such that $\sup_{x \in \mathbf{R}^N} (1 + |x|^2)^{-r/2} |f(x)| < \infty$.

Then we see that $Q_{s,t}^{(L)}$ and $P_{s,t}$, $t > s \geqq 0$, can be regarded as linear operators on $\mathcal{B}_r$.

Now let $\phi_{s,t} : \mathbf{R}^n \times \mathbf{R}$, $s, t \in [0, \infty)$, $s < t$, be measurable functions. We assume that there is a $\lambda \geqq 0$, such that

$$|\phi_{s,t}(x, y) - \phi_{s,t}(x, z)| \leqq \exp(\lambda(t-s))|y-z|, \quad x \in \mathbf{R}^N, y, z \in \mathbf{R}, t > s \geqq 0.$$

Also, we assume that $\phi_{s,t}(\cdot, 0) \in \mathcal{B}_r$, $t > s \geqq 0$.

Let us define a nonlinear operator $\Phi_{s,t} : \mathcal{B}_r \to \mathcal{B}_r$, $s, t \in [0, \infty)$, $s < t$, by

$$(\Phi_{s,t} f)(x) = \phi_{s,t}(x, f(x)), \qquad x \in E, \ f \in \mathcal{B}_r.$$

Then we have

$$N_s^{(L)}(\Phi_{s,t} f - \Phi_{s,t} g) \leqq \exp(\lambda(t-s)) N_s^{(L)}(f - g)$$

for any $f, g \in \mathcal{B}_r$.

Let us define operators $\tilde{Q}_{s,t}^{(L)}$ and $\tilde{P}_{s,t}$ on $\mathcal{B}_r$ by $\tilde{Q}_{s,t}^{(L)} = \Phi_{s,t} \circ Q_{s,t}^{(L)}$ and $\tilde{P}_{s,t} = \Phi_{s,t} \circ P_{s,t}$.

Then we have the following easily from Propositions 14 and 15.

Proposition 16. (1)

$$N_s^{(L)}(\tilde{Q}_{s,t}^{(L)} f - \tilde{Q}_{s,t}^{(L)} g) \leqq \exp(\lambda(t-s)) N_t^{(L)}(f - g)$$

for any $f, g \in \mathcal{B}_r$.

(2) *Let* $T > 0$ *and* $m \geqq 1$. *Then there is a* $C > 0$ *such that*

$$\begin{aligned} &E[N_s^{(L)}(\tilde{Q}_{s,t}^{(L)} f - \tilde{P}_{s,t} f)^2] \\ &\leqq \frac{C}{L} a(L) \exp(2\lambda(t-s))(t-s)^{-(N+1)\ell_0/2} \\ &\times \int_E f(y)^2 (1 + |y|^2)^{-(r+N)} \, dy \end{aligned}$$

for any $L \geqq 1$ *and* $s, t \in [0, T]$ *with* $s < t$. *Here*

$$a(L) = \sup_{s \in [0,T]} E[\max_{\ell = 1, \ldots, L} h(X_\ell(s))^{-2(N+1)} (1 + |X_\ell(s)|^2)^m]$$

Note that by Proposition 15(2), we see that for any $\delta > 0$,

$$L^{-\delta} a(L) \to 0, \qquad L \to \infty.$$

So we have the following.

Theorem 17. *For $T > 0$, there is a $C > 0$ satisfying the following. For any $n \geqq 1$, and $0 = t_0 < t_1 < \cdots < t_n \leqq T$,*

$$E[|(\tilde{Q}^{(L)}_{t_0,t_1} \cdots \tilde{Q}^{(L)}_{t_{n-1},t_n} f)(x_0) - (\tilde{P}_{t_0,t_1} \cdots \tilde{P}_{t_{n-1},t_n} f)(x_0)|^2]^{1/2}$$
$$\leqq \frac{C}{L^{1/2}} a(L)^{1/2} \exp(\lambda t_n) \sum_{k=1}^{n} (t_k - t_{k-1})^{-(N+1)\ell_0/4}$$
$$\times \Big(\int_E (\tilde{P}_{t_k,t_{k+1}} \cdots \tilde{P}_{t_{n-1},t_n} f)(y)^2 (1+|y|^2)^{-(r+N)}\, dy\Big)^{1/2}$$

Proof. Note that

$$|(\tilde{Q}^{(L)}_{t_0,t_1} \cdots \tilde{Q}^{(L)}_{t_{n-1},t_n} f)(x_0) - (\tilde{P}_{t_0,t_1} \cdots \tilde{P}_{t_{n-1},t_n} f)(x_0)|$$
$$= N^{(L)}_0 ((\tilde{Q}^{(L)}_{t_0,t_1} \cdots \tilde{Q}^{(L)}_{t_{n-1},t_n} f) - (\tilde{P}_{t_0,t_1} \cdots \tilde{P}_{t_{n-1},t_n} f))$$
$$\leqq \sum_{k=1}^{n} N^{(L)}_0 ((\tilde{Q}^{(L)}_{t_0,t_1} \cdots \tilde{Q}^{(L)}_{t_{k-1},t_k} \tilde{P}_{t_k,t_{k+1}} \cdots \tilde{P}_{t_{n-1},t_n} f)$$
$$- (\tilde{Q}^{(L)}_{t_0,t_1} \cdots \tilde{Q}^{(L)}_{t_{k-2},t_{k-1}} \tilde{P}_{t_{k-1},t_k} \cdots \tilde{P}_{t_{n-1},t_n} f))$$
$$\leqq \sum_{k=1}^{n} \exp(\lambda t_{k-1}) N^{(L)}_{t_{k-1}} (\tilde{Q}^{(L)}_{t_{k-1},t_k} \tilde{P}_{t_k,t_{k+1}} \cdots \tilde{P}_{t_{n-1},t_n} f)$$
$$- (\tilde{P}_{t_{k-1},t_k} \cdots \tilde{P}_{t_{n-1},t_n} f)).$$

Also, we have by Proposition 16

$$E[N^{(L)}_{t_{k-1}} (\tilde{Q}^{(L)}_{t_{k-1},t_k} \tilde{P}_{t_k,t_{k+1}} \cdots \tilde{P}_{t_{n-1},t_n} f) - (\tilde{P}_{t_{k-1},t_k} \cdots \tilde{P}_{t_{n-1},t_n} f))^2]^{1/2}$$
$$\leqq \frac{C^{1/2}}{L^{1/2}} a(L)^{1/2} \exp(\lambda (t_k - t_{k-1}))(t_k - t_{k-1})^{-(N+1)\ell_0/4}$$
$$\times \Big(\int_E (\tilde{P}_{t_k,t_{k+1}} \cdots \tilde{P}_{t_{n-1},t_n} f)(y)^2 (1+|y|^2)^{-(r+N)}\, dy\Big)^{1/2}.$$

These imply our theorem. ■

Now we apply the above theorem to American option. Let $g : [0,T] \times \mathbf{R}^n \to \mathbf{R}$ be a continuous function such that there are $r \geqq 1$ and $C_1 > 0$ such that $|g(t,x)| \leqq C_1(1+|x|^2)^{r/2}$, $t \in [0,T]$, $x \in \mathbf{R}^n$. Let $\phi_{s,t}(x,y) = g(s,x) \vee y$, for $x \in \mathbf{R}^n$, $y \in \mathbf{R}$, and $s, t \in [0,T]$ with $s < t$. Then we have $\phi_{s,t}(x,y) - \phi_{s,t}(x,z)| \leqq |y - z|$. It is easy to see that there is a $a \geqq 0$ such that

$$E[\sup_{t \in [0,T]} (1+|X(t,x)|^2)^{r/2}] \leqq \exp(aT)(1+|x|^2)^{r/2}, \qquad x \in \mathbf{R}^n.$$

So we see that

$$\sup_{x\in\mathbf{R}^n}(1+|x|^2)^{-r/2}|\tilde{P}_{s,t}f(x)| \leqq \exp C_1 \vee \exp(a(t-s)) \\ \times \sup_{x\in\mathbf{R}^n}(1+|x|^2)^{-r/2}|f(x)|, \quad f\in\mathcal{B}_r.$$

Then we see that

$$(\int_E (\tilde{P}_{t_k,t_{k+1}}\cdots\tilde{P}_{t_{n-1},t_n}g(t_n,\cdot))(y))^2(1+|y|^2)^{-(r+N)}\,dy)^{1/2} \\ \leqq C_1\exp(a(t_n-t_k))\int_E(1+|y|^2)^{-N}\,dy)^{1/2}.$$

So we have by Theorem 17, we see that there is a $C_2>0$ such that

$$E[|(\tilde{Q}^{(L)}_{t_0,t_1}\cdots\tilde{Q}^{(L)}_{t_{n-1},t_n}g(t_n,\cdot)(x_0)-(\tilde{P}_{t_0,t_1}\cdots\tilde{P}_{t_{n-1},t_n}g(t_n,\cdot))(x_0)|^2]^{1/2} \\ \leqq \frac{C_2}{L^{1/2}}a(L)^{1/2}\sum_{k=1}^n(t_k-t_{k-1})^{-(N+1)\ell_0/4}$$

for any $n\geqq 1$, and $0=t_0<t_1<\cdots<t_n\leqq T$. So if we take $n_L\geqq 1$ and $0=t_0^{(L)}<t_1^{(L)}<\cdots<t_n^{(L)}=T$ for each $L\geqq 1$, and there is a $\delta_0,\delta_1>0$, with $\delta_0<\delta_1<1/2$ such that

$$\lim_{L\to\infty}L^{-\delta_0}\sum_{k=1}^{n_L}(t_k^{(L)}-t_{k-1}^{(L)})^{-(N+1)\ell_0/4}=0,$$

then we see that

$$L^{-(1-\delta_1)/2}|(\tilde{Q}^{(L)}_{t_0^{(L)},t_1^{(L)}}\cdots\tilde{Q}^{(L)}_{t_{n_L-1}^{(L)},t_{n_L}^{(L)}}g(T,\cdot)(x_0) \\ -(\tilde{P}_{t_0^{(L)},t_1^{(L)}}\cdots\tilde{P}_{t_{n_L-1}^{(L)},t_{n_L}^{(L)}}g(T,\cdot))(x_0)|\to 0$$

in probability.

5. Preparations for Estimates of Functions

Proposition 18. *Let Z_k, $k=1,2\ldots$ be independent integrable random variables.*

(1) *For any $p\geqq 1$, there is a $C>0$ only depend on p such that*

$$E[|\sum_{k=1}^n(Z_k-E[Z_k])|^{2p}]\leqq C(E[(\sum_{k=1}^n Z_k^2)^p]+(\sum_{k=1}^n|E[Z_k]|)^{2p}),\ n\geqq 1.$$

(2) *For any* $p \geqq 1$, *there is a* $C > 0$ *only depend on* p *such that*

$$E[|\sum_{k=1}^{n} Z_k|^{2p}] \leqq C(E[(\sum_{k=1}^{n} Z_k^2)^p] + (\sum_{k=1}^{n} |E[Z_k]|)^{2p}), \ n \geqq 1.$$

(3) *For any* $m \in \mathbf{N}$, *there is a* $C > 0$ *only depend on* m *such that*

$$E[|\sum_{k=1}^{n} Z_k^2|^{2^m}] \leqq C \sum_{r=1}^{m+1} (\sum_{k=1}^{n} E[Z_k^{2^r}])^{2^{m+1-r}}, \ n \geqq 1.$$

Proof. (1) If $\sum_{k=1}^{n} E[|Z_k|^{2p}] = \infty$, the right hand side is infinity, and so the inequality is valid. So we assume that $\sum_{k=1}^{n} E[|Z_k|^{2p}] < \infty$. Then by Burkholder's inequality we have

$$E[|\sum_{k=1}^{n} (Z_k - E[Z_k])|^{2p}] \leqq C_{2p} E[(\sum_{k=1}^{n} (Z_k - E[Z_k])^2)^p].$$

Since we have

$$E[(\sum_{k=1}^{n} (Z_k - E[Z_k])^2)^p] \leqq 2^p E[(\sum_{k=1}^{n} (Z_k^2 + E[Z_k]^2))^p]$$
$$\leqq 2^{2p} E[(\sum_{k=1}^{n} Z_k^2)^p] + 2^{2p} (\sum_{k=1}^{n} E[Z_k]^2)^p$$
$$\leqq 2^{2p} E[(\sum_{k=1}^{n} Z_k^2)^p] + 2^{2p} (\sum_{k=1}^{n} |E[Z_k]|)^{2p},$$

we have our assertion.

(2) Note that

$$E[|\sum_{k=1}^{n} Z_k|^{2p}] = E[|\sum_{k=1}^{n} ((Z_k - E[Z_k]) + E[Z_k])|^{2p}]$$

$$\leqq 2^{2p} (E[|\sum_{k=1}^{n} (Z_k - E[Z_k]|^{2p}] + |\sum_{k=1}^{n} E[Z_k]|^{2p}).$$

So we have our assertion by the assertion (1).

We can show the assertion (3) easily by induction and the assertion (2). ∎

Proposition 19. *For any $m \geqq 1$, $j \geqq 0$, $\alpha \in \mathbf{Z}^N_{\geqq 0}$, $\delta \in (0,1)$, and $T > 0$, there is a $C > 0$ such that*

$$E[\sup_{s\in[0,t-\varepsilon]} |(\frac{1}{L}\sum_{\ell=1}^{L} \partial_t^j \partial_y^\alpha p(t-s, X_\ell(s), y)) - \partial_t^j \partial_y^\alpha p(t, x_0, y)|^{2^{m+1}}]$$
$$\leqq C\varepsilon^{-2^m(j+|\alpha|+3)\ell_0} L^{-2^m} Lp(t, x_0, y)^{1-\delta}(L^{-1} + p(t, x_0, y)^{1-\delta})^{2^m},$$

for any $y \in \mathbf{R}^N$, $L \geqq 1$, $t \in (0, T]$, $\varepsilon \in (0, t)$.

Proof. Let us note that

$$\frac{\partial}{\partial t}\partial_t^j \partial_y^\alpha p(t, x, y) = L_x \partial_t^j \partial_y^\alpha p(t, x, y), \qquad t > 0,\ x \in E,\ y \in \mathbf{R}^N,$$

where

$$L_x = \frac{1}{2}\sum_{k=1}^{d} V_k^2 + V_0.$$

So we see that $\partial_t^j \partial_y^\alpha p(t-s, X_\ell(s), y)$, $s \in [0, t)$, $h > 0$, is a martingale, and

$$\langle \partial_t^j \partial_y^\alpha p(t-\cdot, X_\ell(\cdot), y)\rangle_s$$
$$= \sum_{k=1}^{d} \int_0^s |\partial_t^j \partial_y^\alpha V_{k,x} p(t-r, X_\ell(r), y)|^2 dr.$$

So we have by Burkholder's inequality and Proposition 18(3),

$$E[\sup_{s\in[0,t-\varepsilon]} |\sum_{\ell=1}^{L}(\partial_t^j \partial_y^\alpha p(t-s, X_\ell(s), y) - \partial_t^j \partial_y^\alpha p(t, x_0, y))|^{2^{m+1}}]$$
$$\leqq C_{2^{m+1}} E[(\sum_{\ell=1}^{L}\sum_{k=1}^{d}\int_0^{t-\varepsilon} |\partial_t^j \partial_y^\alpha V_{k,x} p(t-s, X_\ell(s), y)|^2 ds)^{2^m}]$$
$$\leqq C_{2^{m+1}} d^{2^m} \sum_{k=1}^{d} E[(\sum_{\ell=1}^{L}\int_0^{t-\varepsilon} |\partial_t^j \partial_y^\alpha V_{k,x} p(t-s, X_\ell(s), y)|^2 ds)^{2^m}]$$
$$\leqq C\sum_{k=1}^{d}\sum_{r=0}^{m}(\sum_{\ell=1}^{L} E[(\int_0^{t-\varepsilon} |\partial_t^j \partial_y^\alpha V_{k,x} p(t-s, X_\ell(s), y)|^2 ds)^{2^r}])^{2^{m-r}}$$
$$\leqq C\sum_{k=1}^{d}\sum_{r=0}^{m} t^{2^m - 2^{m-r}}(\sum_{\ell=1}^{L} E[(\int_0^{t-\varepsilon} |\partial_t^j \partial_y^\alpha V_{k,x}$$
$$\times p(t-s, X_\ell(s), y)|^{2^{r+1}} ds])^{2^{m-r}}$$

$$= C\sum_{k=1}^{d}\sum_{r=0}^{m} t^{2^m-2^{m-r}} L^{2^{m-r}} \left(\int_0^{t-\varepsilon} \left(\int_{\mathbf{R}^N} |\partial_t^j \partial_y^\alpha V_{k,x} p(t-s,z,y)|^{2^{r+1}}\right.\right.$$

$$\left.\left. \times\, p(s,x_0,z)dz\right)ds\right)^{2^{m-r}}.$$

Then by Proposition 10, we have

$$E[\sup_{s\in[0,t-\varepsilon]} |\sum_{\ell=1}^{L}(\partial_t^j \partial_y^\alpha p(t-s,X_\ell(s),y) - \partial_t^j\partial_y^\alpha p(t,x_0,y))|^{2^{m+1}}]$$

$$\leqq C' t^{2^m} \varepsilon^{-2^m(j+|\alpha|+3)\ell_0} \sum_{r=0}^{m} L^{2^{m-r}} p(t,x_0,y)^{2^{m-r}(1-\delta)}.$$

$$\leqq C' t^{2^m} \varepsilon^{-2^m(j+|\alpha|+3)\ell_0} L^{2^m} L p(t,x_0,y)^{1-\delta}(L^{-1}+p(t,x_0,y)^{1-\delta})^{2^m}.$$

This implies our assertion. ■

Proposition 20. *For any* $\delta \in (0,1/2)$, $T>0$ *and* $p\in[2,\infty)$, *there is a* $C>0$ *such that*

$$E[(\sup_{y\in\mathbf{R}^N,\, t\in[\varepsilon,T], s\in[0,t-\varepsilon]} \left(\frac{|q_{s,t}^{(L)}(y) - p(t,x_0,y)|}{(L^{-1/(1-\delta)}+p(t,x_0,y))^{(1-\delta)/2}}\right)^p.]^{1/p}$$

$$\leqq C\varepsilon^{-5\ell_0} L^{-1/2+1/p}, \qquad L \geqq 1,\ \varepsilon\in(0,1).$$

Proof. Let us take an $m\geqq 1$ such that $p+N<2^m$. Note that

$$L^{-1}+p(t,x_0,y)^{1-\delta} \leqq 2(L^{-1/(1-\delta)}+p(t,x_0,y))^{1-\delta}.$$

Let

$$\rho_L(s,t,y) = \frac{q_{s,t}^{(L)}(y) - p(t,x_0,y)}{(L^{-1/(1-\delta)}+p(t,x_0,y))^{(1-\delta)/2}}, \quad 0\leqq s<t\leqq T,\ y\in\mathbf{R}^N.$$

We see by Proposition 9, we see that for any $a>0$, $j\geqq 0$, and $\alpha\in\mathbf{Z}^N_{\geqq 0}$, there is a $C>0$ such that

$$(L^{-1/(1-\delta)}+p(t,x_0,y))^{-a+2j+|\alpha|}|\partial_t^j\partial_y^\alpha((L^{-1/(1-\delta)}+p(t,x_0,y))^{-a})|$$

$$\leqq Ct^{-(2j+|\alpha|)\ell_0} p(t,x_0,y)^{-\delta}, \qquad y\in\mathbf{R}^N,\ t\in(0,T].$$

So we see that by Proposition 18, for any $a>0$, $j=0,1$, and $\alpha\in\mathbf{Z}^N_{\geqq 0}$ with $|\alpha|\leqq 1$, there is a $C>0$ such that

$$E[\sup_{s\in[0,t-\varepsilon]} |\partial_t^j \partial_y^\alpha \rho_L(s,t,y)|^{2^{m+1}}]$$
$$\leqq C\varepsilon^{-2^{m+1}4\ell_0} L^{-2^m} Lp(t,x_0,y)^{1-2\delta},$$
$$y \in \mathbf{R}^N,\ L \geqq 1,\ \varepsilon \in (0,1),\ t \in (\varepsilon, T].$$

Therefore we see that

$$E[\int_{\mathbf{R}^N} dy \sup_{s\in[0,t-\varepsilon]} |\partial_t^j \partial_y^\alpha \rho_L(s,t,y)|^{2^{m+1}}]$$
$$\leqq C\varepsilon^{-2^{m+3}\ell_0} L^{-2^m} L \int_{\mathbf{R}^N} p(t,x_0,y)^{1-2\delta} dy,$$
$$L \geqq 1,\ \varepsilon \in (0,1),\ t \in (\varepsilon, T].$$

Note by Proposition 8 that there is a $C > 0$ such that

$$\int_{\mathbf{R}^N} p(t,x_0,y)^{1-2\delta} dy \leqq Ct^{(N+1)\ell_0\delta}, \qquad t \in (0,T].$$

Also, note that

$$\partial_y^\alpha \rho_L(s,t,y) = \partial_y^\alpha \rho_L(s,T,y) - \int_t^T \partial_r \partial_y^\alpha \rho_L(s,r,y) dr,$$

and so we see that

$$\sup_{t\in[\varepsilon,T], s\in[0,t-\varepsilon]} \int_{\mathbf{R}^N} |\partial_y^\alpha \rho_L(s,t,y)|^{2^m} dy$$
$$\leqq 2^{m+1} \int_{\mathbf{R}^N} dy \sup_{s\in[0,T-\varepsilon]} |\partial_y^\alpha \rho_L(s,T,y)|^{2^m}$$
$$+ 2^{m+1}(T+1)^{2^m} \int_t^T dr \int_{\mathbf{R}^N} dy \sup_{s\in[0,r-\varepsilon)} |\partial_y^\alpha \rho_L(s,r,y)|^{2^m}.$$

Then by Sobolev's inequality, we see that there is a $C > 0$ such that

$$E[\sup_{y\in\mathbf{R}^N} \sup_{t\in[\varepsilon,T], s\in[0,t-\varepsilon]} |\rho_L(s,t,y)|^{2^{m+1}}]^{1/2^{m+1}}$$
$$\leqq C\varepsilon^{-(4+(N+1)/2^{m+1})\ell_0} L^{-1/2+1/2^{m+1}},$$

$L \geqq 1,\ \varepsilon \in (0,1)$. This implies our assertion. ■

Let

$$Z_L(s,t;\delta) = \sup_{y\in\mathbf{R}^N} \frac{|q_{s,t}^{(L)}(y) - p(t,x_0,y)|}{(L^{-1/(1-\delta)} + p(t,x_0,y))^{(1-\delta)/2}}, \qquad t>0,\ s\in[0,t)$$

and

$$\tilde{Z}_L(\varepsilon,\delta,T) = \sup_{t\in[\varepsilon,T],s\in[0,t-\varepsilon]} Z_L(s,t;\delta)$$

for $T > 0$, $\varepsilon \in (0,T]$, and $\delta \in (0,1)$. Note that $Z_L(s,t;\delta)$ is $\mathcal{F}_s^{(\infty)}$-measurable.

Then we have the following.

Proposition 21. (1) *Let $T > 0$, $\varepsilon \in (0,T]$, and $\delta \in (0,1)$. Then for any $p > 1$, there is a $C > 0$ such that*

$$E[(L^{(1-\delta^2)/2}\tilde{Z}_L(\varepsilon,\delta,T))^p]^{1/p} \leqq C\varepsilon^{-5\ell_0}L^{-p\delta^2/2+1/p}, \qquad L\geqq 1.$$

(2) *Let $\delta \in (0,1)$, $t>0$, and $s\in(0,t)$. If $L^{(1-\delta^2)/2}Z_L(s,t;\delta) \leqq 1/4$, and $p(t,x_0,y) \geqq L^{-(1-\delta)}$, then*

$$\frac{1}{2} \leqq \frac{q_{s,t}^{(L)}(y)}{p(t,x_0,y)} \leqq 2, \qquad t\in(\varepsilon,T],\ s\in[0,t-\varepsilon].$$

Proof. The assertion (1) is an immediate consequence of Proposition 20. Note that

$$|q_{s,t}^{(L)}(y) - p(t,x_0,y)| \leqq Z_L(s,t;\delta)(L^{-1/(1-\delta)} + p(t,x_0,y))^{(1-\delta)/2}$$

for any $y\in\mathbf{R}^N$, $t\in[\varepsilon,T]$ and $s\in[0,t-\varepsilon]$.

If $p(t,x_0,y) \geqq L^{-(1-\delta)}$, we have

$$\left|\frac{q_{s,t}^{(L)}(y)}{p(t,x_0,y)} - 1\right| \leqq Z_L(s,t;\delta)(L^{-1/(1-\delta)}p(t,x_0,y)^{-1} + 1)^{(1-\delta)/2}$$
$$\times\, p(t,x_0,y)^{-(1+\delta)/2}$$
$$\leqq Z_L(s,t;\delta)(L^{-1/(1-\delta)}L^{1-\delta}+1)^{(1-\delta)/2}L^{(1-\delta^2)/2} \leqq 2L^{(1-\delta^2)/2}Z_L(s,t;\delta).$$

This implies our second assertion. ■

Proposition 22. *Let $T > 0$, and $\delta\in(0,1)$. Let $B_L(s,t)\in\mathcal{F}$, $L\geqq 1$, be given by*

$$B_L(s,t) = \{\omega\in\Omega:\ L^{(1-\delta^2)/2}Z_L(s,t;\delta) \leqq 1/4\}, \qquad t > and\, s\in(0,t),$$

and $\varphi_{t,L}: E\to\{0,1\}$, $t\in(0,T]$, $L\geqq 1$, be given by

$$\varphi_{t,L} = 1_{\{y\in E;\, p(t,x_0,y)>L^{-(1-\delta)}\}}, \qquad t>0.$$

(1) *Let* $a \in (1/(2N), 1/2)$, $b \in (a - 1/(2N), a)$, *and* $m \geqq 1$. *Then there is a* $C > 0$ *such that*

$$1_{B_L(s,t)} E[(\sup_{x \in E} p(s, x_0, x)^a |(Q_{s,t}^{(L)}(\varphi_{t,L} f))(x) - (P_{s,t}(\varphi_{t,L} f))(x)|)^2 | \mathcal{F}_s^{(\infty)}] \leqq \frac{C}{L} s^{-(N+2)\ell_0} (t-s)^{-(N+2)\ell_0} \int_E p(t, x_0, y)^{-1+2b} (1 + |y|^2)^{-m} \times \varphi_{t,L}(y) f(y)^2 dy \quad a.s.$$

for $t \in (0, T]$, $s \in (0, t)$, $L \geqq 1$, *and any bounded measurable function* f *defined in* E.

(2) *Let* $a \in (0, 1/2)$, *and* $m \geqq 1$. *Then there is a* $C > 0$ *such that*

$$1_{B_L(s,t)} E[(\sup_{x \in E} p(s, x_0, x)^{1/2-\delta/4} |Q_{s,t}^{(L)}(\varphi_{t,L} f))(x) - (P_{s,t}(\varphi_{t,L} f))(x)|)^2 | \mathcal{F}_s^{(\infty)}] \leqq \frac{C}{L^{1-\delta}} s^{-(N+2)\ell_0} (t-s)^{-(N+2)\ell_0} \times \int_E (1 + |y|^2)^{-m} \varphi_{t,L}(y) f(y)^2 dy \quad a.s.$$

for $t \in (0, T]$, $s \in (0, t)$, $L \geqq 1$, *and any bounded measurable function* f *defined in* E.

(3) *Let* $a \in (0, 1/2)$ *and* $b \in (a - 1/(2N), a)$. *Then there is a* $C > 0$ *such that*

$$1_{B_L(s,t)} E[(\sup_{x \in E} p(s, x_0, x)^a |Q_{s,t}^{(L)}(\varphi_{t,L} p(t, x_0, \cdot)^{-b}))(x) - (P_{s,t}(\varphi_{t,L} p(t, x_0, \cdot)^{-b}))(x)|)^2 | \mathcal{F}_s^{(\infty)}] \leqq \frac{C}{L^{\delta}} s^{-(N+2)\ell_0} (t-s)^{-(N+2)\ell_0} \quad a.s.$$

for $t \in (0, T]$, $s \in (0, t)$, $L \geqq 1$.

Proof. Note that for $\alpha \in \mathbf{Z}_{\geqq 0}^N$

$$\begin{aligned} &1_{B_L(s,t)} E[|\partial_x^\alpha (p(s, x_0, x)^a (Q_{s,t}^{(L)}(\varphi_{t,L} f))(x) - (P_{s,t}(\varphi_{t,L} f))(x)))|^2 \mathcal{F}_s^{(\infty)}] \\ &\leqq \frac{1}{L} 1_{B_L(s,t)} \int_E \frac{|\partial_x^\alpha (p(s, x_0, x)^a p(t-s, x, y))|^2}{q_{s,t}^{(L)}(y)} \varphi_{t,L(y)} f(y)^2 \, dy \\ &\leqq \frac{2}{L} 1_{B_L(s,t)} \int_E |\partial_x^\alpha (p(s, x_0, x)^a p(t-s, x, y))|^2 \\ &\quad \times p(t, x_0, y)^{-1} \varphi_{t,L(y)} f(y)^2 \, dy \end{aligned}$$

So we have by Proposition 12 there is a $C > 0$ such that

$$1_{B_L(s,t)}E[\int_{\mathbf{R}^N} dx\ |\partial_x^\alpha(p(s,x_0,x)^a(Q_{s,t}^{(L)}(\varphi_{t,L}f))(x) - (P_{s,t}(\varphi_{t,L}f))(x)))|^2\mathcal{F}_s^{(\infty)}]$$

$$\leqq \frac{C}{L}s^{-(N+2)\ell_0}(t-s)^{-(N+2)\ell_0}\int_E p(t,x_0,y)^{-1+2b}(1+|y|^2)^{-m}\varphi_{t,L}(y)f(y)^2dy.$$

This and Sobolev's inequality imply the assertion (1).

In the assertion (1), if $a = 1 - \delta/4$ and $b > 1/2 - \delta/2$, then we have

$$p(t,x_0,y)^{-1+2b}\varphi_{t,L}(y) \leqq L^{-\delta}.$$

This implies the assertion (2).

In the assertion (1), if $m = N + 1$ and $f = p(t,x_0,\cdot)^{-b}$ then we have

$$\int_E p(t,x_0,y)^{-1+2b}(1+|y|^2)^{-m}\varphi_{t,L}(y)f(y)^2dy \leqq L^{1-\delta}\int_{\mathbf{R}^N}(1+|y|^2)^{-(N+1)}dy$$

This implies the assertion (3). ■

Similarly by using Proposition 12, we have the following.

Proposition 23. *Let $a \in (1/(2N), 1/2)$ and $b \in (a - 1/(2N), a)$. Then there is a $C > 0$ such that*

$$\sup_{x\in E} p(s,x_0,x)^a|(P_{s,t}f)(x)|$$
$$\leqq Cs^{-(N+2)\ell_0/2}(t-s)^{-(N+3)\ell_0/2}\sup_{y\in E} p(t,x_0,y)^b|f(y)|$$

for $t \in (0,T]$, $s \in (0,t)$, and any bounded measurable function f defined in E.

6. Application to Bermuda Type Problem

Let us think of the situation in Sect. 4. Then we have the following.

Theorem 24. *Let $0 = T_0 < T_1 < \ldots < T_n < T$, $\delta \in (0,1/2)$, and $f \in \mathcal{B}_r$, for some $r \geqq 0$. Then there are $C > 0$, $\Omega^L \in \mathcal{F}$, $L \geqq 1$, and measurable functions $d_{m,i}^{(L)} : E \times \Omega \to [0,\infty)$, $m = 1,\ldots,n-1$, $i = 1,2$, $L \geqq 1$, such that*

$$\lim_{L\to\infty} L^p(1-P(\Omega^L))=0, \qquad p\in(1,\infty),$$

$$1_{\Omega^L}|(\tilde{Q}^{(L)}_{T_m,T_{m+1}}\cdots\tilde{Q}^{(L)}_{T_{n-1},T_n}f)(x)-(\tilde{P}_{T_m,T_{m+1}}\cdots\tilde{P}_{T_{n-1},T_n}f)(x)|$$
$$\leqq d^{(L)}_{m,1}(x)+d^{(L)}_{m,2}(x), \qquad x\in E,\ m=1,\ldots,n-1,\ L\geqq 1$$

and

$$E[\int_E d^{(L)}_{m,1}(x)p(T_m,x_0,x)dx]\leqq CL^{-(1-\delta)^2}$$
$$E[\int_E d^{(L)}_{m,2}(x)^2p(T_m,x_0,x)dx]\leqq CL^{-(1-\delta)}$$

for any $L\geqq 1$, $m=1,\ldots,n-1$.

Proof. Note that for $f,g\in\mathcal{B}_{r'}$

$$|(\tilde{Q}^{(L)}_{s,t}f)(x)-(\tilde{Q}^{(L)}_{s,t}g)(x)|$$
$$=|\phi_{s,t}(x,(Q^{(L)}_{s,t}f)(x))-\phi_{s,t}(x,(Q^{(L)}_{s,t}g)(x))|$$
$$\leqq \exp(\lambda(t-s))(Q^{(L)}_{s,t}(|f-g|))(x)$$

So we see that

$$|(\tilde{Q}^{(L)}_{T_m,T_{m+1}}\cdots\tilde{Q}^{(L)}_{T_{k-1},T_k}f)(x)-(\tilde{Q}^{(L)}_{T_m,T_{m+1}}\cdots\tilde{Q}^{(L)}_{T_{k-1},T_k}g)(x)|$$
$$\leqq \exp(\lambda(T_k-T_m))(Q_{T_m,T_{m+1}}\cdots Q^{(L)}_{T_{k-1}.T_k}(|f-g|))(x)$$

Similarly we have

$$|(\tilde{Q}^{(L)}_{s,t}f)(x)-(\tilde{P}_{s,t}g)(x)|\leqq \exp(\lambda(t-s))|(Q^{(L)}_{s,t}f)(x)-(P_{s,t}g)(x)|$$

Let us take a_k, $k=0,1,\ldots,n$ such that $1/2>a_0>a_1>\ldots>a_n>1/2-\delta$. Also, let

$$c_m(x)=(\tilde{P}_{T_m,T_{m+1}}\cdots\tilde{P}_{T_{n-1},T_n}f)(x).$$

Note that

$$|(\tilde{Q}^{(L)}_{T_m,T_{m+1}}\cdots\tilde{Q}^{(L)}_{T_{n-1},T_n}f)(x)-(\tilde{P}_{T_m,T_{m+1}}\cdots\tilde{P}_{T_{n-1},T_n}f)(x)|$$
$$\leqq\sum_{k=1}^{n-m}|(\tilde{Q}^{(L)}_{T_m,T_{m+1}}\cdots\tilde{Q}^{(L)}_{T_{m+k-1},T_{m+k}}\tilde{P}_{T_{m+k},T_{m+k+1}}\cdots\tilde{P}_{T_{n-1},T_n}f)(x)$$
$$-(\tilde{Q}^{(L)}_{T_m,T_{m+1}}\cdots\tilde{Q}^{(L)}_{T_{m+k-2},T_{m+k-1}}\tilde{P}_{T_{m+k-1},T_{m+k}}\cdots\tilde{P}_{T_{n-1},T_n}f)(x)|$$

$$\leqq \exp(\lambda T) \sum_{k=1}^{n-m} (Q^{(L)}_{T_m,T_{m+1}} \cdots Q^{(L)}_{T_{m+k-2},T_{m+k-1}} (|Q^{(L)}_{T_{m+k-1},T_{m+k}} c_{m+k}$$
$$- P_{T_{m+k-1},T_{m+k}} c_{m+k}|))(x).$$

Let

$$R_k = 1_{B_L(T_{k-1},T_k)} \sup_{x \in E} p(T_{k-1}, x_0, x)^{a_{k-1}} (|Q^{(L)}_{T_{k-1},T_k}(\varphi_{T_k,L} c_k)$$
$$- P_{T_{k-1},T_k}(\varphi_{T_k,L} c_k)|)(x),$$
$$Z_k = 1_{B_L(T_{k-1},T_k)} \sup_{x \in E} p(T_{k-1}, x_0, x)^{a_{k-1}}$$
$$\times (|Q^{(L)}_{T_{k-1},T_k}(\varphi_{T_k,L} p(T_k, x_0, \cdot)^{-a_k})$$
$$- P_{T_{k-1},T_k}(\varphi_{T_k,L} p(T_k, x_0, \cdot)^{-a_k})|)(x),$$

and

$$D_k = \sup_{x \in E} p(T_{k-1}, x_0, x)^{a_{k-1}} (P_{T_{k-1},T_k}(\varphi_{T_k,L} p(T_k, x_0, \cdot)^{-a_k}))(x) < \infty,$$
$$k = 1, \ldots, n.$$

Then R_k and Z_k are $\mathcal{F}^{(\infty)}_{T_k}$-measurable for $k = 1, \ldots, n$, and by Proposition 22 we see that there is a $C > 0$ such that

$$E[R_k^2 | \mathcal{F}^{(\infty)}_{T_{k-1}}] \leqq C L^{-1}, \qquad E[Z_k^2 | \mathcal{F}^{(\infty)}_{T_{k-1}}] \leqq C L^{-(1-\delta)}$$

for any $L \geqq 1$, and $k = 1, \ldots, n$. So inductively we have

$$E[R_k^2 (\prod_{i=\ell+1}^{k} (Z_i + D_i)^2) | \mathcal{F}^{(\infty)}_{T_\ell}] \leqq 2^{k-\ell} C^{k+1-\ell} L^{-1} \prod_{i=\ell+1}^{k} (D_i^2 + C L^{-(1-\delta)})$$

for any $L \geqq 1$, and $1 \leqq \ell \leqq k \leqq n$. Let $\Omega^L = \bigcap_{k=1}^n B_L(T_{k-1}, T_k)$. Then we have

$$1_{\Omega^L} Q^{(L)}_{T_{k-1},T_k}(p(T_k, x_0, \cdot)^{-a_k})(x)$$
$$= 1_{\Omega^L} Q^{(L)}_{T_{k-1},T_k}(\varphi_{T_k,L} p(T_k, x_0, \cdot)^{-a_k})(x)$$
$$+ Q^{(L)}_{T_{k-1},T_k}((1 - \varphi_{T_k,L}) p(T_k, x_0, \cdot)^{-a_k})(x)$$
$$\leqq 1_{\Omega^L} (Z_k + D_k) p(T_{k-1}, x_0, x)^{-a_{k-1}}$$
$$+ 1_{\Omega^L} Q^{(L)}_{T_{k-1},T_k}((1 - \varphi_{T_k,L}) p(T_k, x_0, \cdot)^{-a_k})(x).$$

Therefore we have

$$1_{\Omega^L}(Q^{(L)}_{T_m,T_{m+1}}\cdots Q^{(L)}_{T_{m+k-2},T_{m+k-1}}(|Q^{(L)}_{T_{m+k-1},T_{m+k}}c_{m+k}$$
$$- P_{T_{m+k-1},T_{m+k}}c_{m+k}|)(x) \leqq 1_{\Omega^L}R_{m+k}(Q^{(L)}_{T_m,T_{m+1}}\cdots Q^{(L)}_{T_{m+k-2},T_{m+k-1}}$$
$$\times p(T_{m+k-1},x_0,\cdot)^{-a_{m+k-1}})(x)$$
$$+ 1_{\Omega^L}(Q^{(L)}_{T_m,T_{m+1}}\cdots Q^{(L)}_{T_{m+k-2},T_{m+k-1}}(Q^{(L)}_{T_{m+k-1},T_{m+k}}((1-\varphi_{T_{m+k},L})|c_{m+k}|)$$
$$+ P_{T_{m+k-1},T_{m+k}}((1-\varphi_{T_{m+k},L})|c_{m+k}|)))(x)$$
$$\leqq \tilde{d}_{m,2}(x) + \tilde{d}_{m.1}(x),$$

where

$$\tilde{d}^{(L)}_{m,2}(x) = R_{m+k}(\prod_{i=1}^{k}(Z_{m+i} + D_{m+i}))p(T_m,x_0,x)^{-a_m})$$

and

$$\tilde{d}^{(L)}_{m,1}(x) = \sum_{\ell=1}^{k} R_{m+k}(\prod_{i=m+\ell+1}^{m+k}(Z_i + D_i))$$
$$\times (Q^{(L)}_{T_m,T_{m+1}}\cdots Q^{(L)}_{T_{m+\ell-2},T_{m+\ell-1}}((1-\varphi_{T_{m+\ell-1},L})p(T_{m+\ell-1},x_0,\cdot)^{a_{m+\ell-1}}))(x))$$
$$+ (Q^{(L)}_{T_m,T_{m+1}}\cdots Q^{(L)}_{T_{m+k-2},T_{m+k-1}}(Q^{(L)}_{T_{m+k-1},T_{m+k}}((1-\varphi_{T_{m+k},L})|c_{m+k}|)$$
$$+ P_{T_{m+k-1},T_{m+k}}((1-\varphi_{T_{m+k},L})|c_{m+k}|)))(x).$$

Note that

$$E[R_{m+k}(\prod_{i=m+\ell+1}^{m+k}(Z_i + D_i))(Q^{(L)}_{T_m,T_{m+1}}\cdots Q^{(L)}_{T_{m+\ell-2},T_{m+\ell-1}}((1-\varphi_{T_{m+\ell-1},L})$$
$$\times p(T_{m+\ell-1},x_0,\cdot)^{a_{m+\ell-1}})(x))]$$
$$= E[E[R_{m+k}(\prod_{i=m+\ell+1}^{m+k}(Z_i + D_i))|\mathcal{F}^{(\infty)}_{T_{m+\ell}}]$$
$$\times (Q^{(L)}_{T_m,T_{m+1}}\cdots Q^{(L)}_{T_{m+\ell-2},T_{m+\ell-1}}((1-\varphi_{T_{m+\ell-1},L})$$
$$\times p(T_{m+\ell-1},x_0,\cdot)^{a_{m+\ell-1}})(x))]$$
$$\leqq (2^kC^{k+1}L^{-1}\prod_{i=m+\ell+1}^{m+k}(D_i^2 + CL^{-(1-\delta)}))^{1/2}$$
$$\times E[Q^{(L)}_{T_m,T_{m+1}}\cdots Q^{(L)}_{T_{m+\ell-2},T_{m+\ell-1}}((1-\varphi_{T_{m+\ell-1},L})$$

$$\times p(T_{m+\ell-1}, x_0, \cdot)^{a_{m+\ell-1}})(x))]$$

$$= (2^k C^{k+1} L^{-1} \prod_{i=m+\ell+1}^{m+k} (D_i^2 + CL^{-(1-\delta)}))^{1/2} P_{T_m, T_{m+\ell-1}}((1 - \varphi_{T_{m+\ell-1}, L})$$

$$\times p(T_{m+\ell-1}, x_0, \cdot)^{a_{m+\ell-1}})(x).$$

Note that for $a \geqq 0$

$$\int_E P_{T_m, T_{m+\ell-1}}((1 - \varphi_{T_{m+\ell-1}, L}) p(T_{m+\ell-1}, x_0, \cdot)^a)(x) p(T_m, x_0, x) dx$$

$$= \int_E 1_{\{p(T_{m+\ell-1}, x_0, x) \leqq L^{-(1-\delta)}\}} p(T_{m+\ell-1}, x_0, x)^{1+a} dx$$

$$\leqq L^{-(1-\delta)^2} \int_E p(T_{m+\ell-1}, x_0, x)^{\delta + a} dx.$$

Then we have our assertion. ■

7. Re-simulation

We think of application to pricing Bermuda derivatives.

Let $r \geqq 1$ and let $g : [0, T] \times \mathbf{R}^N \to \mathbf{R}$ be a continuous function such that

$$\sup_{x \in \mathbf{R}^N,\, t \in [0,T]} (1 + |x|^2)^{-r/2} |g(t, x)| < \infty.$$

Let $\phi_{s,t}(x, y) = g(s, x) \vee y$, $0 \leqq s < t \leqq T$, $x \in \mathbf{R}^N$ and $y \in \mathbf{R}$. Let $0 = T_0 < T_1 < \ldots < T_n < T$, and let $c_m : E \to \mathbf{R}$, $m = 0, 1, \ldots, n$, be given by

$$c_m(x) = (\tilde{P}_{T_m, T_{m+1}} \cdots \tilde{P}_{T_{n-1}, T_n} g(T_n, \cdot))(x), \; m \leqq n - 1, \text{ and } c_n(x) = g(T_n.x).$$

Now let $\tilde{c}_m : E \to \mathbf{R}$, $m = 1, \ldots, n - 1$, be given and let $\tilde{c}_n = g(T_n, \cdot)$. We regard $\tilde{c}_m$ as estimators of c_m, $m = 1, \ldots, n$.

Let us think of the SDE in Introduction. Let $\tau : W_0 \to \{T_1, \ldots, T_n\}$ and $\tilde{\tau} : W_0 \to \{T_1, \ldots, T_n\}$ be stopping times given by

$$\tau = \min\{T_k;\; c_k(X(T_k, x_0)) \leqq g(T_k, X(T_k, x_0)),\; k = 1, \ldots, n\}$$

and

$$\tilde{\tau} = \min\{T_k;\; \tilde{c}_k(X(T_k, x_0)) \leqq g(T_k, X(T_k, x_0)),\; k = 1, \ldots, n\}$$

Let $\bar{c}_m$, $m = 0, \dots, n$, be given by inductively, $\bar{c}_n = g(T_n, \cdot)$, and

$$\bar{c}_{m-1} = P_{T_{m-1},T_m}(g(T_m, \cdot)1_{\{\tilde{c}_m \leqq g(T_m,\cdot)\}} + \bar{c}_m 1_{\{\tilde{c}_m > g(T_m,\cdot)\}}), \quad m = n, n-1, \dots, 1.$$

Then we have the following.

Proposition 25. (1) *For* $m = 0, 1, \dots, n-1$,

$$E^\mu[g(\tau, X(\tau, x_0)|\mathcal{B}_{T_m}]1_{\{\tau \geqq T_{m+1}\}} = c_m(X(T_m, x_0))1_{\{\tau \geqq T_{m+1}\}} \ a.s.$$

and

$$E^\mu[g(\tilde{\tau}, X(\tilde{\tau}, x_0)|\mathcal{B}_{T_m}]1_{\{\tilde{\tau} \geqq T_{m+1}\}} = \bar{c}_m(X(T_m, x_0))1_{\{\tilde{\tau} \geqq T_{m+1}\}} \ a.s.$$

Here $\mathcal{B}_t = \sigma\{B^i(s);\ s \leqq t,\ i = 1, \dots, d\}$.
(2) *For* $m = 0, 1, \dots, n-1$, *and* $x \in E$,

$$0 \leqq c_m(x) - \bar{c}_m(x) \leqq P_{T_m,T_{m+1}}(|c_{m+1} - \bar{c}_{m+1}|)(x)$$
$$+ P_{T_m,T_{m+1}}(1_{\{\tilde{c}_{m+1} > g_{m+1}\}}(c_{m+1} - \bar{c}_{m+1}))(x).$$

In particular,

$$0 \leqq c_m(x) - \bar{c}_m(x) \leqq \sum_{k=m+1}^{n} P_{T_m,T_k}(|c_k - \bar{c}_k|)(x), \qquad m = 0, 1, \dots, n.$$

Proof. Since we have

$$\begin{aligned} &E^\mu[g(\tilde{\tau}, X(\tilde{\tau}, x_0)|\mathcal{B}_{T_{m-1}}]1_{\{\tilde{\tau} \geqq T_m\}} \\ &= E^\mu[E^\mu[g(\tilde{\tau}, X(\tilde{\tau}, x_0)1_{\{\tilde{\tau} \geqq T_{m+1}\}}|\mathcal{B}_{T_m}] \\ &\quad + g(T_m, X(T_m, x_0))1_{\{\tilde{\tau} = T_m\}}|\mathcal{B}_{T_{m-1}}], \end{aligned}$$

we can easily obtain the assertion (1) by induction.

Note that

$$\begin{aligned} &c_m - \bar{c}_m \\ &= P_{T_m,T_{m+1}}(1_{\{\tilde{c}_{m+1} \leqq g(T_{m+1},\cdot)\}}((g(T_{m+1}, \cdot) \vee c_{m+1}) - g(T_{m+1}, \cdot))) \\ &\quad + P_{T_m,T_{m+1}}(1_{\{\tilde{c}_{m+1} > g(T_{m+1},\cdot)\}}((g(T_{m+1}, \cdot) \vee c_{m+1}) - \bar{c}_{m+1})) \\ &= P_{T_m,T_{m+1}}(1_{\{\tilde{c}_{m+1} \leqq g(T_{m+1},\cdot)\}}((g(T_{m+1}, \cdot) \vee c_{m+1}) \\ &\quad - (g(T_{m+1}, \cdot) \vee \bar{c}_{m+1}))) + P_{T_m,T_{m+1}}(1_{\{\tilde{c}_{m+1} > g(T_{m+1},\cdot)\}}((g(T_{m+1}, \cdot) \\ &\quad - c_{m+1}) \vee 0) - ((g(T_{m+1}, \cdot) - \bar{c}_{m+1}) \vee 0) + c_{m+1} - \bar{c}_{m+1})) \\ &\leqq P_{T_m,T_{m+1}}(|c_{m+1} - \bar{c}_{m+1}|) + P_{T_m,T_{m+1}}(1_{\{\tilde{c}_{m+1} > g(T_{m+1},\cdot)\}}(c_{m+1}) - \bar{c}_{m+1})) \end{aligned}$$

This implies the first inequality of the assertion (2). The second inequality follows from this by induction. ■

Proposition 26.

$$
\begin{aligned}
&c_0(x_0) - \bar{c}_0(x_0) \\
&\leqq \sum_{k=1}^{n} \int_E (|\tilde{c}_k - \bar{c}_k| + |c_k - \bar{c}_k|)(x) 1_{\{|\tilde{c}_k - c_k| + |c_k - \bar{c}_k| \geqq \varepsilon\}}(x) \\
&\quad + \varepsilon 1_{\{|g(T_k,\cdot) - c_k| < \varepsilon\}} p(T_k, x_0, x) dx
\end{aligned}
$$

for any $\varepsilon > 0$.

Proof. Note that

$$
\begin{aligned}
c_0(x_0) - \bar{c}_0(x_0) &= E^{\mu}[g(\tau, X(\tau, x_0)) - g(\tilde{\tau}, X(\tilde{\tau}, x_0))] \\
&= E^{\mu}[g(\tau, X(\tau, x_0)) - g(\tilde{\tau}, X(\tilde{\tau}, x_0)), \tau > \tilde{\tau}] + E^{\mu}[g(\tau, X(\tau, x_0)) \\
&\quad - g(\tilde{\tau}, X(\tilde{\tau}, x_0)), \tau < \tilde{\tau}] \\
&= E^{\mu}[E^{\mu}[g(\tau, X(\tau, x_0))|\mathcal{B}_{\tilde{\tau}}] - g(\tilde{\tau}, X(\tilde{\tau}, x_0), \tau > \tilde{\tau}] \\
&\quad + E^{\mu}[g(\tau, X(\tau, x_0) - E^{\mu}[g(\tilde{\tau}, X(\tilde{\tau}, x_0))|\mathcal{B}_{\tau}], \tau < \tilde{\tau}] \\
&= \sum_{k=1}^{n-1} (E^{\mu}[c_k(X(T_k, x_0)) - g(T_k, X(T_k, x_0)), \tau > T_k, \tilde{\tau} = T_k] \\
&\quad + E^{\mu}[g(T_k, X(k, x_0)) - \tilde{c}_k(X(T_k, x_0)), \tau = T_k, T_k < \tilde{\tau}]) \\
&\leqq \sum_{k=1}^{n-1} (E^{\mu}[(c_k(X(T_k, x_0)) - g(T_k, X(T_k, x_0)) 1_{\{\tilde{c}_k \leqq g(T_k,\cdot) < c_k\}}(X(T_k, x_0))] \\
&\quad + E^{\mu}[((g(T_k, X(k, x_0)) - \tilde{c}_k(X(T_k, x_0)))) \vee 0) \\
&\quad \times 1_{\{c_k \leqq g(T_k,\cdot) < \tilde{c}_k\}}(X(T_k, x_0))]).
\end{aligned}
$$

For any $\varepsilon > 0$, we see that

$$
\begin{aligned}
&(c_k - g(T_k, \cdot)) 1_{\{\tilde{c}_k \leqq g(T_k,\cdot) < c_k\}} \\
&\quad \leqq \varepsilon 1_{\{g(T_k,\cdot) < c_k \leqq g(T_k,\cdot) + \varepsilon\}} + (c_k - g(T_k, \cdot)) 1_{\{\tilde{c}_k \leqq g(T_k,\cdot) < c_k\}} 1_{\{g_k + \varepsilon < c_k\}} \\
&\quad \leqq \varepsilon 1_{\{g(T_k,\cdot) < (T_k,\cdot) + \varepsilon\}} + (c_k - \tilde{c}_k) 1_{\{c_k - \tilde{c}_k > \varepsilon\}} 1_{\{\tilde{c}_k \leqq g(T_k,\cdot) < c_k\}},
\end{aligned}
$$

and

$$
\begin{aligned}
&((g(T_k, \cdot) - \tilde{c}_k) \vee 0) 1_{\{c_k \leqq g(T_k,\cdot) < \tilde{c}_k\}} \\
&\quad \leqq ((g(T_k, \cdot) - \tilde{c}_k) \vee 0) 1_{\{c_k \leqq g(T_k,\cdot) < \tilde{c}_k\}} 1_{\{|\tilde{c}_k - c_k| + |c_k - \bar{c}_k| \geqq \varepsilon\}} \\
&\qquad + ((g(T_k, \cdot) - \tilde{c}_k) \vee 0) 1_{\{c_k \leqq g(T_k,\cdot) < \tilde{c}_k\}} 1_{\{|\tilde{c}_k - c_k| + |c_k - \bar{c}_k| < \varepsilon\}} \\
&\quad \leqq (|\tilde{c}_k - c_k| + |c_k - \bar{c}_k|) 1_{\{|\tilde{c}_k - c_k| + |c_k - \bar{c}_k| \geqq \varepsilon\}} 1_{\{c_k \leqq g(T_k,\cdot) < \tilde{c}_k\}} \\
&\qquad + \varepsilon 1_{\{c_k \leqq g(T_k,\cdot) < c_k + \varepsilon\}}.
\end{aligned}
$$

So we have our assertion. ■

Now we have the following.

Lemma 27. *Let $d_{m,i} : E \to [0,\infty)$. $m = 1, \dots, n$, $i = 1, 2$, be measurable functions. Assume that $|\tilde{c}_m - c_m| \leqq d_{m,1} + d_{m,2}$, $m = 1, \dots, n$. Then we have the following.*

$$
\begin{aligned}
&c_0(x_0) - \bar{c}_0(x_0) \\
&\leqq n \sum_{k=1}^{n} \int_E d_{k,1}(x) p(T_k, x_0, x) dx + n(\sum_{k=1}^{n} (\int_E d_{k,2}(x)^2 p(T_k, x_0, x) dx)^{1/2}) \\
&\quad \times (\varepsilon^{-1/2} (\sum_{k=1}^{n} \int_E d_{k,1}(x) p(T_k, x_0, x) dx) \\
&\quad + \varepsilon^{-1} (\sum_{k=1}^{n} (\int_E d_{k,2}(x)^2 p(T_k, x_0, x) dx)^{1/2})) \\
&\quad + 2\varepsilon \sum_{k=1}^{n} \int_E 1_{\{|g(T_k,\cdot) - c_k| < 2\varepsilon\}} p(T_k, x_0, x) dx
\end{aligned}
$$

for any $\varepsilon > 0$.

Proof. Let

$$
\tilde{d}_{m,i}(x) = \sum_{k=m}^{n} (P_{T_m,T_k} d_{k,i})(x), \quad m = 1, \dots, n.
$$

Then by Proposition 25, we have

$$
|\tilde{c}_m(x) - c_m(x)| + |\bar{c}_m(x) - c_m(x)| \leqq \tilde{d}_{m,1}(x) + \tilde{d}_{m,2}(x).
$$

Note that

$$
\begin{aligned}
&\int_E (\tilde{d}_{m,1}(x) + \tilde{d}_{m,2}(x)) 1_{\{\tilde{d}_{m,1}(x) + \tilde{d}_{m,2}(x) \geqq 2\varepsilon\}}(x) p(T_m, x_0, x) dx \\
&\quad \leqq \int_E \tilde{d}_{m,1}(x) p(T_m, x_0, x) dx + (\int_E \tilde{d}_{m,2}(x)^2 p(T_m, x_0, x) dx)^{1/2} \\
&\quad\quad \times ((\int_E 1_{\{\tilde{d}_{m,1}(x) \geqq \varepsilon\}}(x) p(T_m, x_0, x) dx)^{1/2} \\
&\quad\quad + (\int_E 1_{\{\tilde{d}_{m,2}(x) \geqq \varepsilon\}}(x) p(T_m, x_0, x) dx)^{1/2}) \\
&\quad \leqq \int_E \tilde{d}_{m,1}(x) p(T_m, x_0, x) dx + (\int_E \tilde{d}_{m,2}(x)^2 p(T_m, x_0, x) dx)^{1/2}
\end{aligned}
$$

$$\times (\varepsilon^{-1/2}(\int_E \tilde{d}_{m,1}(x)p(T_m, x_0, x)dx)^{1/2}$$
$$+ \varepsilon^{-1}(\int_E \tilde{d}_{m,2}(x)^2 p(T_m, x_0, x)dx)^{1/2})$$

Also, note that

$$\int_E \tilde{d}_{m,1}(x)p(T_m, x_0, x)dx \leqq \sum_{k=m}^{n} \int_E d_{k,1}(x)p(T_k, x_0, x)dx,$$

and

$$(\int_E \tilde{d}_{m,2}(x)^2 p(T_m, x_0, x)dx)^{1/2} \leqq \sum_{k=m}^{n} (\int_E d_{k,2}(x)^2 p(T_k, x_0, x)dx)^{1/2}.$$

This and Proposition 26 imply our assertion. ■

Now we apply this lemma and the results in the previous section to a Bermuda derivative.

Let $\phi_{s,t}(x, y) = g(s, x) \vee y$, $0 \leqq s < t \leqq T$, $x \in \mathbf{R}^N$ and $y \in \mathbf{R}$. Let $\tilde{c}_m : E \to \mathbf{R}$, $m = 1, \ldots, n-1$, be given by

$$\tilde{c}_m(x) = (\tilde{Q}^{(L)}_{T_m,T_{m+1}} \cdots \tilde{Q}^{(L)}_{T_{n-1},T_n} g(T_n, \cdot))(x).$$

Then by Theorem 24, we see that for any $\delta \in (0, 1/2)$, there are $\Omega'_L \in \mathcal{F}$, $L \geqq 1$, $C > 0$ and measurable functions $d^{(L)}_{m,i} : E \times \Omega \to [0, \infty)$, $m = 1, \ldots, n-1$, $i = 1, 2$, $L \geqq 1$, such that

$$\lim_{L\to\infty} P(\Omega'_L) = 1,$$

$$|\tilde{c}_m(x) - c_m(x)| \leqq d_{m,1}(x) + d_{m,2}(x), \quad x \in E,\ \omega \in \Omega'_L,\ m = 1, \ldots, n-1,\ L \geqq 1$$

and

$$E[\int_E d_{m,1}(x)p(T_m, x_0, x)dx] \leqq CL^{-(1-\delta)^2} \qquad m = 1, \ldots, n-1,\ L \geqq 1,$$

and

$$E[\int_E d_{m,2}(x)^2 p(T_m, x_0, x)dx] \leqq CL^{-(1-\delta)^2} \quad m = 1, \ldots, n-1,\ L \geqq 1.$$

Let

$$\Omega_L'' = \{\omega \in \Omega; \int_E d_{m,1}(x) p(T_m, x_0, x) dx \geqq L^{-(1-\delta)^3} \text{ or }$$
$$\int_E d_{m,2}(x)^2 p(T_m, x_0, x) dx \geqq L^{-(1-\delta)^3}\}.$$

Then we see that

$$P(\Omega \setminus \Omega_L'') \leqq 2CL^{-(1-\delta)^2\delta}, \qquad L \geqq 1.$$

Let $\Omega_L = \Omega_L' \cap \Omega_L''$, $L \geqq 1$. Then we see that $P(\Omega_L) \to 1$, $l \to \infty$. So if we use these $\tilde{c}_m(x)$, $m = 1, \ldots, n-1$, as estimators and use the re-simulation method, we have

$$\begin{aligned} &c_0(x_0) - \bar{c}_0(x_0) \\ &\leqq n^2 L^{-(1-\delta)^3} + n^3 L^{-(1-\delta)^3/2} (\varepsilon^{-1/2} L^{-(1-\delta)^3/2} + \varepsilon^{-1} L^{-(1-\delta)^3/2}) \\ &\quad + \varepsilon \sum_{k=1}^{n} \int_E 1_{\{|g(T_k, \cdot) - c_k| < \varepsilon\}} p(T_k, x_0, x) dx \end{aligned}$$

for any $\varepsilon > 0$, $\omega \in \Omega_L$, and $L \geqq 1$. Suppose that

$$\sum_{k=1}^{n-1} \int_E 1_{\{|g(T_k, \cdot) - c_k| < \varepsilon\}} p(T_k, x_0, x) dx = O(\varepsilon^\gamma), \qquad \varepsilon \downarrow 0,$$

for some $\gamma \in (0, 1]$. Then letting $\varepsilon = L^{-(1-\delta)^3/(2+\gamma)}$, we see that $c_0(x_0) - \bar{c}_0(x_0) = O(L^{-(1-\delta)^3(1+\gamma)/(2+\gamma)})$ as $L \to \infty$.

Since δ is arbitrary, this proves Theorem 2.

References

1. Avramidis, A.N., Hyden, P.: Efficiency improvements for pricing American options with a stochastic mesh. In: Proceedings of the 1999 Winter Simulation Conference, pp. 344–350 (1999)
2. Avramidis, A.N., Matzinger, H.: Convergence of the stochastic mesh estimator for pricing American options. J. Comput. Finance **7**(4), 73–91 (2004)
3. Belomestny, D.: Pricing Bermudan options by nonparametric regression: optimal rates of convergence for lower estimates. Finance Stoch. **15**, 655–683 (2011)

4. Broadie, M., Glasserman, P.: A stochastic mesh method for pricing high-dimensional American options. J. Comput. Finance **7**(4), 35–72 (2004)
5. Glasserman, P.: Monte Carlo Methods in Financial Engineering. Springer, Berlin (2004)
6. Kusuoka, S.: Malliavin calculus revisited. J. Math. Sci. Univ. Tokyo **10**, 261–277 (2003)
7. Kusuoka, S., Stroock, D.W.: Applications of Malliavin calculus II. J. Fac. Sci. Univ. Tokyo Sect. IA Math. **32**, 1–76 (1985)
8. Shigekawa, I.: Stochastic Analysis. Translation of Mathematical Monographs, vol. 224. AMS, Providence (2000)

Adv. Math. Econ. 18, 101–134 (2014)

Advances in
MATHEMATICAL
ECONOMICS

Turnpike Properties for Nonconcave Problems

Alexander J. Zaslavski

Department of Mathematics, The Technion – Israel Institute of Technology,
Technion City, Haifa 32000, Israel
(e-mail: ajzasl@techunix.technion.ac.il)

Received: November 12, 2013
Revised: November 28, 2013

JEL classification: C6, O4

Mathematics Subject Classification (2010): 49J99, 91B55, 91B62

Abstract. In this survey paper we discuss recent developments in the turnpike theory for nonconvex (nonconcave) problems. We also establish a new result on agreeable solutions for variational problems with extended-valued integrands.

Key words: Agreeable function, Good program, Overtaking optimal program, Turnpike property

1. Introduction

The study of the existence and the structure of solutions of optimal control problems defined on infinite intervals and on sufficiently large intervals has recently been a rapidly growing area of research. See, for example, [2, 3, 6–10, 13, 14, 16, 18, 22, 23, 25, 26, 29, 31, 36–38, 40, 51, 57, 59] and the references mentioned therein. These problems arise in engineering [1, 27, 60], in models of economic growth [4, 11, 12, 17, 24, 30, 35, 39, 42–44, 51], in infinite discrete models of solid-state physics related to dislocations in one-dimensional crystals [5, 45] and in the theory of thermodynamical equilibrium for materials [15, 28, 32–34]. In this paper we discuss the structure of solutions of a discrete-time optimal control system describing a general model of economic dynamics and the structure of solutions of variational problems with extended-valued integrands. We study an autonomous discrete-time control system with a compact metric space of states X. This control system is described by a bounded upper semicontinuous function $v : X \times X \to R^1$ which determines an optimality criterion and by a nonempty closed set $\Omega \subset X \times X$ which determines a class of admissible trajectories (programs). In models of economic growth the set X is the space of states, v is a utility function and $v(x_t, x_{t+1})$ evaluates consumption at moment t.

We are interested in a turnpike property of the approximate solutions which is independent of the length of the interval, for all sufficiently large intervals. To have this property means, roughly speaking, that approximate solutions of optimal control problems on an interval $[0, T]$ with given values y, z at the endpoints 0 and T, corresponding to the pair (v, Ω), are determined mainly by the objective function v, and are essentially independent of T, y and z. Turnpike properties are well known in mathematical economics. The term was first coined by Samuelson in 1948 (see [43]) where he showed that an efficient expanding economy would spend most of the time in the vicinity of a balanced equilibrium path (also called a von Neumann path). This property was further investigated for optimal trajectories of models of economic dynamics (see, for example, [24, 30, 35, 42, 51] and the references mentioned there). In the classical turnpike theory the function v has the turnpike property (TP) if there exists $\bar{x} \in X$ (a turnpike) which satisfies the following condition:

For each $\epsilon > 0$ there is a natural number L such that for each integer $T \geq 2L$ and each solution $\{x_i\}_{i=0}^T \subset X$ of an optimal control problem corresponding to the pair (v, Ω), for all $i = L, \dots, T - L$, the point x_i belongs to an ϵ-neighborhood of $\bar{x}$.

Note that L does not depend on T.

In the classical turnpike theory the space X is a compact convex subset of a finite-dimensional Euclidean space, the set Ω is convex and the function v is strictly concave. Under these assumptions the turnpike property can be established and the turnpike $\bar{x}$ is a unique solution of the maximization problem $v(x, x) \to \max$, $(x, x) \in \Omega$.

In this situation it is shown that for each admissible sequence $\{x_t\}_{t=0}^\infty$ satisfying $(x_t, x_{t+1}) \in \Omega$ for all integers $t \geq 0$, either the sequence $\{\sum_{t=0}^{T-1} v(x_t, x_{t+1}) - Tv(\bar{x}, \bar{x})\}_{T=1}^\infty$ is bounded (in this case the sequence $\{x_t\}_{t=0}^\infty$ is called (v)-good) or it diverges to $-\infty$. Moreover, it is also established that any (v)-good admissible sequence converges to the turnpike $\bar{x}$. In the sequel this property is called as the asymptotic turnpike property.

More recently, it was shown that the turnpike property is a general phenomenon which holds for large classes of variational and optimal control problems without convexity assumptions [46–49, 51]. For these classes of problems a turnpike is not necessarily a singleton but may instead be an nonstationary trajectory (in the discrete time nonautonomous case) or an absolutely continuous function on the interval $[0, \infty)$ (in the continuous time nonautonomous case) or a compact subset of the space X (in the autonomous case) [51]. Nevertheless, problems for which the turnpike is a singleton are of great importance because of the following reasons: there are many models of economic growth for which a turnpike is a singleton; if a turnpike is a singleton, then approximate solutions have very simple structure and this is very important for applications; if a turnpike is a singleton, then it can be easily calculated as a solution of the problem $v(x, x) \to \max$, $(x, x) \in \Omega$.

The results presented in the paper explain when the turnpike property holds with the turnpike being a singleton. We show that the turnpike property follows from the asymptotic turnpike property. More precisely, we assume that any (v)-good admissible trajectory converges to a unique solution $\bar{x}$ of the problem $v(x, x) \to \max$, $(x, x) \in \Omega$ and show that the turnpike property holds and $\bar{x}$ is the turnpike. Note that we do not use convexity (concavity) assumptions.

The paper is organized as follows. In Sect. 2 we describe the class of discrete-time optimal control problems. For this class of problems we present turnpike results and show the existence of optimal solutions over infinite horizon. Section 3 contains two auxiliary results. In Sect. 4 we discuss the structure of solutions of the discrete-time problems in the regions containing end points. In Sect. 5 we study turnpike properties of approximate solutions of variational problems with extended-valued integrands. For this class of variational problems we also show the existence of optimal solutions over infinite horizon. Section 6 contains two auxiliary propositions. Examples of variational problems are considered in Sect. 7. In Sect. 8 we discuss the structure of solutions of the variational problems with extended-valued integrands in the regions containing end points. In Sect. 9 we show equivalence of optimality criterions used in the literature for our class of variational problems. A new result on agreeable solutions is presented in Sect. 10. Its proof is given in Sect. 11.

2. Discrete-Time Problems

Let (X, ρ) be a compact metric space, Ω be a nonempty closed subset of $X \times X$ and let $v : X \times X \to R^1$ be a bounded upper semicontinuous function.

A sequence $\{x_t\}_{t=0}^{\infty} \subset X$ is called an (Ω)-program (or just a program if the set Ω is understood) if $(x_t, x_{t+1}) \in \Omega$ for all nonnegative integers t. A sequence $\{x_t\}_{t=T_1}^{T_2}$ where integers T_1, T_2 satisfy $0 \leq T_1 < T_2$ is called an (Ω)-program (or just a program if the set Ω is understood) if $(x_t, x_{t+1}) \in \Omega$ for all integers $t \in [T_1, T_2 - 1]$.

We consider the problems

$$\sum_{i=0}^{T-1} v(x_i, x_{i+1}) \to \max, \ \{(x_i, x_{i+1})\}_{i=0}^{T-1} \subset \Omega, \ x_0 = y, \ x_T = z, \quad (P_T^{(y,z)})$$

and

$$\sum_{i=0}^{T-1} v(x_i, x_{i+1}) \to \max, \ \{(x_i, x_{i+1})\}_{i=0}^{T-1} \subset \Omega, \ x_0 = y, \qquad (P_T^{(y)})$$

where $T \geq 1$ is an integer and the points $y, z \in X$.

Set $\|v\| = \sup\{|v(x, y)| : x, y \in X\}$. For each pair of points $x, y \in X$ and each natural number T define

$$\sigma(v, T, x) = \sup\{\sum_{i=0}^{T-1} v(x_i, x_{i+1}) : \{x_i\}_{i=0}^{T} \text{ is a program and } x_0 = x\}, \tag{2.1}$$

$$\sigma(v, T, x, y) = \sup\{\sum_{i=0}^{T-1} v(x_i, x_{i+1}) : \{x_i\}_{i=0}^{T} \text{ is a program and } x_0 = x, \ x_T = y\}, \tag{2.2}$$

$$\sigma(v, T) = \sup\{\sum_{i=0}^{T-1} v(x_i, x_{i+1}) : \{x_i\}_{i=0}^{T} \text{ is a program}\}. \tag{2.3}$$

(Here we use the convention that the supremum of an empty set is $-\infty$).

We suppose that there exist a point $\bar{x} \in X$ and a positive constant $\bar{c}$ such that the following assumptions hold.

(A1) $(\bar{x}, \bar{x})$ is an interior point of Ω (there exists a positive number ϵ such that $\{(x, y) \in X \times X : \rho(x, \bar{x}), \ \rho(y, \bar{x}) \leq \epsilon\} \subset \Omega$) and the function v is continuous at the point $(\bar{x}, \bar{x})$.

(A2) $\sigma(v, T) \leq T v(\bar{x}, \bar{x}) + \bar{c}$ for all natural numbers T.

Clearly, for each integer $T \geq 1$ and each program $\{x_t\}_{t=0}^{T}$ we have

$$\sum_{t=0}^{T-1} v(x_t, x_{t+1}) \leq \sigma(v, T) \leq T v(\bar{x}, \bar{x}) + \bar{c}. \tag{2.4}$$

Inequality (2.4) implies the following result.

Proposition 2.1. *For each program* $\{x_t\}_{t=0}^{\infty}$ *either the sequence*

$$\{\sum_{t=0}^{T-1} v(x_t, x_{t+1}) - Tv(\bar{x}, \bar{x})\}_{T=1}^{\infty}$$

is bounded or $\lim_{T\to\infty}[\sum_{t=0}^{T-1} v(x_t, x_{t+1}) - Tv(\bar{x}, \bar{x})] = -\infty$.

A program $\{x_t\}_{t=0}^{\infty}$ is called (v)-good if the sequence

$$\{\sum_{t=0}^{T-1} v(x_t, x_{t+1}) - Tv(\bar{x}, \bar{x})\}_{T=1}^{\infty}$$

is bounded.

We suppose that the following assumption holds.

(A3) (the asymptotic turnpike property) For every (v)-good program $\{x_t\}_{t=0}^{\infty}$,

$$\lim_{t\to\infty} \rho(x_t, \bar{x}) = 0.$$

In view of (A3) $\|v\| > 0$. For each positive number M denote by X_M the set of all points $x \in X$ for which there exists a program $\{x_t\}_{t=0}^{\infty}$ such that $x_0 = x$ and that for all natural numbers T the following inequality holds:

$$\sum_{t=0}^{T-1} v(x_t, x_{t+1}) - Tv(\bar{x}, \bar{x}) \geq -M.$$

It is not difficult to see that $\cup\{X_M : M \in (0, \infty)\}$ is the set of all points $x \in X$ such that there exists a (v)-good program $\{x_t\}_{t=0}^{\infty}$ satisfying $x_0 = x$.

Let $T \geq 1$ be an integer. Denote by Y_T the set of all points $x \in X$ such that there exists a program $\{x_t\}_{t=0}^{T}$ which satisfies $x_0 = \bar{x}$ and $x_T = x$.

The following turnpike result describes the structure of approximate solutions of problem $(P_T^{(y)})$.

Theorem 2.2. *Let* ϵ, M *be positive numbers. Then there exist a natural number* L *and a positive number* δ *such that for each integer* $T > 2L$ *and each program* $\{x_t\}_{t=0}^{T}$ *which satisfies*

$$x_0 \in X_M, \sum_{t=0}^{T-1} v(x_t, x_{t+1}) \geq \sigma(v, T, x_0) - \delta \tag{2.5}$$

there exist nonnegative integers $\tau_1, \tau_2 \leq L$ *such that* $\rho(x_t, \bar{x}) \leq \epsilon$ *for all* $t = \tau_1, \ldots, T - \tau_2$ *and if* $\rho(x_0, \bar{x}) \leq \delta$, *then* $\tau_1 = 0$.

In the sequel we use a notion of an overtaking optimal program introduced in [4, 17, 44].

A program $\{x_t\}_{t=0}^{\infty}$ is called (v)-overtaking optimal if for each program $\{y_t\}_{t=0}^{\infty}$ satisfying $y_0 = x_0$ the inequality

$$\limsup_{T\to\infty} \sum_{t=0}^{T-1} [v(y_t, y_{t+1}) - v(x_t, x_{t+1})] \le 0$$

holds.

The following result establishes the existence of an overtaking optimal program.

Theorem 2.3. *Assume that $x \in X$ and that there exists a (v)-good program $\{x_t\}_{t=0}^{\infty}$ such that $x_0 = x$. Then there exists an (v)-overtaking optimal program $\{x_t^*\}_{t=0}^{\infty}$ such that $x_0^* = x$.*

The next theorem is a refinement of Theorem 2.2. According to Theorem 2.2 we have $\tau_2 \le L$ where the constant L depends on M and ϵ. The next theorem shows that $\tau_2 \le L_0$ where the constant L_0 depends only on ϵ.

Theorem 2.4. *Let ϵ be positive number. Then there exists a natural number L_0 such that for each positive number M there exist an integer $L > L_0$ and a positive number δ such that the following assertion holds:*

For each integer $T > 2L$ and each program $\{x_t\}_{t=0}^{T}$ which satisfies (2.5) there exist integers $\tau_1 \in [0, L]$, $\tau_2 \in [0, L_0]$ such that $\rho(x_t, \bar{x}) \le \epsilon$ for all $t = \tau_1, \dots, T - \tau_2$ and if $\rho(x_0, \bar{x}) \le \delta$, then $\tau_1 = 0$.

The following result provides necessary and sufficient conditions for overtaking optimality.

Theorem 2.5. *Let $\{x_t\}_{t=0}^{\infty}$ be a program such that*

$$x_0 \in \cup\{X_M : M \in (0, \infty)\}.$$

Then the program $\{x_t\}_{t=0}^{\infty}$ is (v)-overtaking optimal if and only if the following conditions hold:

(i) $\lim_{t\to\infty} \rho(x_t, \bar{x}) = 0$;
(ii) for each natural number T and each program $\{y_t\}_{t=0}^{T}$ satisfying $y_0 = x_0$, $y_T = x_T$ the inequality $\sum_{t=0}^{T-1} v(y_t, y_{t+1}) \le \sum_{t=0}^{T-1} v(x_t, x_{t+1})$ holds.

The next two theorems establish uniform convergence of overtaking optimal programs to $\bar{x}$.

Theorem 2.6. *Assume that the function v is continuous and let ϵ be a positive number. Then there exists a positive number δ such that for each (v)-overtaking optimal program $\{x_t\}_{t=0}^{\infty}$ satisfying $\rho(x_0, \bar{x}) \le \delta$ the inequality $\rho(x_t, \bar{x}) \le \epsilon$ holds for all nonnegative integers t.*

Theorem 2.7. *Assume that the function v is continuous and let M, ϵ be positive numbers. Then there exists an integer $L \ge 1$ such that for each (v)-overtaking optimal program $\{x_t\}_{t=0}^{\infty}$ satisfying $x_0 \in X_M$ the inequality $\rho(x_t, \bar{x}) \le \epsilon$ holds for all integers $t \ge L$.*

Theorems 2.2–2.7 were obtained in [52]. The next two theorems obtained in [55] describe the structure of problem $(P_T^{(y,z)})$.

Denote by Card(A) the cardinality of the set A.

Theorem 2.8. *Let ϵ, M_0, M_1 be positive numbers and let L_0 be a natural number. Then there exist a natural number L and a natural number K such that for each integer $T > 2L$, each $z_0 \in X_{M_0}$ and each $z_1 \in Y_{L_0}$, $\sigma(v, T, z_0, z_1)$ is finite and for each program $\{x_t\}_{t=0}^{T}$ which satisfies*

$$x_0 = z_1, \ x_T = z_2, \ \sum_{t=0}^{T-1} v(x_t, x_{t+1}) \ge \sigma(v, T, z_0, z_1) - M_1$$

the following inequality holds:

$$Card(\{t \in \{0, \ldots, T\} : \rho(x_t, \bar{x}) > \epsilon\}) \le K.$$

Theorem 2.9. *Let ϵ, M_0 be positive numbers and let L_0 be a natural number. Then there exists a natural number L and a positive number δ such that for each integer $T > 2L$, each $z_0 \in X_{M_0}$ and each $z_1 \in Y_{L_0}$, $\sigma(v, T, z_0, z_1)$ is finite and for each program $\{x_t\}_{t=0}^{T}$ which satisfies*

$$x_0 = z_1, \ x_T = z_2, \ \sum_{t=0}^{T-1} v(x_t, x_{t+1}) \ge \sigma(v, T, z_0, z_1) - \delta$$

there exist integers $\tau_1, \tau_2 \in [0, L]$ such that

$$\rho(x_t, \bar{x}) \le \epsilon, \ t = \tau_1, \ldots, T - \tau_2.$$

Moreover if $\rho(x_0, \bar{x}) \le \delta$, then $\tau_1 = 0$ and if $\rho(x_T, \bar{x}) \le \delta$. then $\tau_2 = 0$.

Example 2.10. Let (X, ρ) be a compact metric space, Ω be a nonempty closed subset of $X \times X$, $\bar{x} \in X$, $(\bar{x}, \bar{x})$ be an interior point of Ω, $\pi : X \to R^1$ be a continuous function, α be a real number and $L : X \times X \to [0, \infty)$

be a continuous function such that for each $(x, y) \in X \times X$ the equality $L(x, y) = 0$ holds if and only if $(x, y) = (\bar{x}, \bar{x})$. Set

$$v(x, y) = \alpha - L(x, y) + \pi(x) - \pi(y)$$

for all $x, y \in X$. It is not difficult to see that (A1), (A2) and (A3) hold.

Example 2.11. Let X be a compact convex subset of the Euclidean space R^n with the norm $|\cdot|$ induced by the scalar product $\langle \cdot, \cdot \rangle$, let $\rho(x, y) = |x - y|$, $x, y \in R^n$, Ω be a nonempty closed subset of $X \times X$, a point $\bar{x} \in X$, $(\bar{x}, \bar{x})$ be an interior point of Ω and let $v : X \times X \to R^1$ be a strictly concave continuous function such that

$$v(\bar{x}, \bar{x}) = \sup\{v(z, z) : z \in X \text{ and } (z, z) \in \Omega\}.$$

We assume that there exists a positive constant $\bar{r}$ such that

$$\{(x, y) \in R^n \times R^n : |x - \bar{x}|, \ |y - \bar{x}| \le \bar{r}\} \subset \Omega.$$

It is a well-know fact of convex analysis that there exists a point $l \in R^n$ such that

$$v(x, y) \le v(\bar{x}, \bar{x}) + \langle l, x - y \rangle$$

for any point $(x, y) \in X \times X$. Set

$$L(x, y) = v(\bar{x}, \bar{x}) + \langle l, x - y \rangle - v(x, y)$$

for all $(x, y) \in X \times X$. It is not difficult to see that this example is a particular case of Example 2.10. Therefore (A1)–(A3) hold.

Examples 2.10 and 2.11 were considered in [52].

3. Two Auxiliary Results

For each integer $T \ge 1$ denote by $\bar{Y}_T$ the set of all points $x \in X$ for which there exists a program $\{x_t\}_{t=0}^T$ such that $x_0 = x$ and $x_T = \bar{x}$.

It is easy to see that

$$\bar{Y}_T \subset \bar{Y}_{T+1} \text{ for all natural numbers } T.$$

By assumption (A1), if T is a natural number and a point $x \in \bar{Y}_T$, then there exists a (v)-good program $\{x_t\}_{t=0}^\infty$ which satisfies $x_0 = x$. Assumptions (A1) and (A3) imply that if a program $\{x_t\}_{t=0}^\infty$ is (v)-good, then there exists an integer $T \ge 1$ such that $x_0 \in \bar{Y}_T$. The boundedness of v implies the following result.

Proposition 3.1. *Let T be a natural number. Then there exists a positive number M such that $\bar{Y}_T \subset X_M$.*

The next result was obtained in [52].

Proposition 3.2. *Let M be positive number. Then there exists an integer $T \geq 1$ such that the inclusion $X_M \subset \bar{Y}_T$ holds.*

In view of assumption (A1), there exists a number $\bar{r} \in (0, 1)$ such that

$$\{(x, y) \in X \times X : \ \rho(x, \bar{x}), \rho(y, \bar{x}) \leq \bar{r}\} \subset \Omega.$$

In view of the inclusion above for each pair of points $x, y \in X$ such that

$$\rho(x, \bar{x}), \rho(y, \bar{x}) \leq \bar{r}$$

and each natural number T, the value $\sigma(v, x, y, T)$ is finite.

4. Structure of Solutions in the Regions Containing End Points

Assume that X is a compact convex subset of the n-dimensional Euclidean space R^n with the norm $|\cdot|$ induced by the scalar product $\langle \cdot, \cdot \rangle$, Ω is a nonempty closed convex subset of $X \times X$ and that $v : X \times X \to R^1$ is a continuous strictly concave function such that

$$v(\alpha z_1 + (1-\alpha)z_2) > \alpha v(z_1) + (1-\alpha)v(z_2)$$

$$\forall \alpha \in (0, 1), \forall z_1, z_2 \in X \times X \text{ such that } z_1 \neq z_2.$$

Put $\rho(x, y) = |x - y|$ for all $x, y \in X$. We assume that $\bar{x} \in X, \bar{r} \in (0, 1)$ and that

$$v(\bar{x}, \bar{x}) = \sup\{v(z, z) : \ z \in X \text{ and } (z, z) \in \Omega)\},$$
$$\{(x, y) \in R^n \times R^n : \ |x - \bar{x}|, \ |y - \bar{x}| \leq 2\bar{r}\} \subset \Omega.$$

We have mentioned (see Example 2.11) that there is $l \in R^n$ such that

$$v(x, y) \leq v(\bar{x}, \bar{x}) + \langle l, x - y \rangle \text{ for all } (x, y) \in X \times X.$$

Set

$$L(x, y) = v(\bar{x}, \bar{x}) + \langle l, x - y \rangle - v(x, y), \ (x, y) \in X \times X.$$

It is clear that the inequality $L(x, y) \geq 0$ holds for all points $x, y \in X$. It was explained in Sect. 2 that assumptions (A1)–(A3) hold. Therefore Theorems 2.2–2.7 hold for the function v. Since the set Ω is convex and the function v is strictly concave Theorem 2.3 implies the following result.

Theorem 4.1. *Assume that $x \in X$ and that there exists a (v)-good program $\{x_t\}_{t=0}^{\infty}$ such that $x_0 = x$. Then there exists a unique (v)-overtaking program $\{x_t^*\}_{t=0}^{\infty}$ such that $x_0^* = x$.*

Let $z \in X$ be given and let there exist a (v)-good program $\{x_t\}_{t=0}^{\infty} \subset X$ such that $x_0 = z$. Denote by $\{x_t^{(v,z)}\}_{t=0}^{\infty}$ a unique (v)-overtaking optimal program satisfying $x_0^{(v,z)} = z$.

The following theorem which describes the structure of approximate solutions in the region containing the left end point of the interval $[0, T]$ was obtained in [52].

Theorem 4.2. *Let $M, \epsilon > 0$ be given and $L_0 \geq 1$ be an integer. Then there exists a positive number δ and an integer $L_1 > L_0$ such that for each natural number $T \geq L_1$ and each program $\{z_t\}_{t=0}^{T}$ which satisfies*

$$z_0 \in X_M, \sum_{t=0}^{T-1} v(z_t, z_{t+1}) \geq \sigma(v, T, z_0) - \delta$$

the inequality $|z_t - x_t^{(v,z_0)}| \leq \epsilon$ holds for all integers $t = 0, \ldots, L_0$.

It follows from Theorem 2.4 applied with $\epsilon = \bar{r}/4$ that there exists a natural number L_0 such that the following property holds:

For each positive number M there exist an integer $L > L_0$ and a positive number δ such that if a natural number $T > 2L$ and if a program $\{x_t\}_{t=0}^{T}$ satisfies

$$x_0 \in X_M, \sum_{t=0}^{T-1} v(x_t, x_{t+1}) \geq \sigma(v, T, x_0) - \delta,$$

then

$$|x_t - \bar{x}| \leq \bar{r}/4 \text{ for all integers } t = L, \ldots, T - L_0.$$

Define the functions $\bar{L}, \bar{v} : X \times X \to R^1$ and the set $\bar{\Omega}$ by

$$\bar{v}(x, y) = v(y, x), \ \bar{L}(x, y) = L(y, x), \ x, y \in X,$$

$$\bar{\Omega} = \{(x, y) \in X \times X : \ (y, x) \in \Omega\}.$$

It is easy to see that $\bar{\Omega}$ is a nonempty closed convex subset of $X \times X$, $\bar{v} : X \times X \to R^1$ is a concave function,

$$\{(\xi_1, \xi_2) \in R^n \times R^n : \ |\xi_i - \bar{x}| \leq 2\bar{r}, \ i = 1, 2\} \subset \bar{\Omega},$$

$$\bar{v}(\bar{x}, \bar{x}) = \sup\{\bar{v}(z, z) : \ z \in X \text{ and } (z, z) \in \bar{\Omega}\},$$

$$\bar{v}(\alpha z_1 + (1-\alpha)z_2) > \alpha\bar{v}(z_1) + (1-\alpha)\bar{v}(z_2) \text{ for each } z_1, z_2 \in X \times X$$
$$\text{satisfying } z_1 \neq z_2 \text{ and each } \alpha \in (0,1).$$

Evidently, for all points $(x, y) \in X \times X$, we have

$$\bar{v}(x, y) \leq \bar{v}(\bar{x}, \bar{x}) + \langle -l, x - y \rangle,$$

$$\bar{L}(x, y) = \bar{v}(\bar{x}, \bar{x}) + \langle -l, x - y \rangle - \bar{v}(x, y).$$

It is not difficult to see that the assumptions (A1)–(A3) hold for the function $\bar{v}$ and the set $\bar{\Omega}$ and that Theorems 2.2–2.7, 4.1 and 4.2 hold for $\bar{v}$ and $\bar{\Omega}$.

Denote by X_* the set of all points $x \in X$ for which there exists an $(\bar{\Omega})$-program $\{x_t\}_{t=0}^{L_0+1}$ such that

$$x_0 = x, \; x_{L_0+1} = \bar{x}.$$

It is clear that X_* is a closed and convex set.

In view of Theorem 4.1 for any point $x \in X_*$ there exists a unique $(\bar{v})$-overtaking optimal $(\bar{\Omega})$-program $\{\Lambda_t(x)\}_{t=0}^{\infty}$ such that $\Lambda_0(x) = x$. For any point $x \in X_*$ put

$$\pi(x) = \lim_{T \to \infty} \sum_{t=0}^{T-1} [\bar{v}(\bar{x}, \bar{x}) - \bar{v}(\Lambda_t(x), \Lambda_{t+1}(x))]$$
$$= \lim_{T \to \infty} [\sum_{t=0}^{T-1} \bar{L}(\Lambda_t(x), \Lambda_{t+1}(x)) + \langle l, x - \Lambda_T(x) \rangle]$$
$$= \sum_{t=0}^{\infty} \bar{L}(\Lambda_t(x), \Lambda_{t+1}(x)) + \langle l, x - \bar{x} \rangle.$$

It is not difficult to show that $\pi(x)$ is finite for all points $x \in X_*$.

In order to study the structure of approximate solutions of the problems $(P_T^{(y)})$ in the regions $[T - L, T]$ (see the definition of the turnpike property) we need the following auxiliary results obtained in [52].

Proposition 4.3. *An $(\bar{\Omega})$-program $\{x_t\}_{t=0}^{\infty}$ is $(\bar{v})$-good if and only if*

$$\sum_{t=0}^{\infty} \bar{L}(x_t, x_{t+1}) < \infty.$$

Proposition 4.4. *Let $x \in X_*$ and let an $(\bar{\Omega})$-program $\{x_t\}_{t=0}^{\infty}$ be $(\bar{v})$-good and satisfy $x_0 = x$. Then $\sum_{t=0}^{\infty} \bar{L}(x_t, x_{t+1}) + \langle l, x - \bar{x} \rangle \geq \pi(x)$.*

Proposition 4.5. *The function $\pi : X_* \to R^1$ is lower semicontinuous.*

Proposition 4.6. *Let points $y, z \in X_*$ satisfy $y \neq z$ and a number $\alpha \in (0, 1)$. Then $\pi(\alpha y + (1-\alpha)z) < \alpha\pi(y) + (1-\alpha)\pi(z)$.*

Since the function $\pi : X_* \to R^1$ is lower semicontinuous and strictly convex it possesses a unique point of minimum which will be denoted by x_*. Thus

$$\pi(x_*) < \pi(x) \text{ for all points } x \in X_* \setminus \{x_*\}.$$

The next theorem which describes the structure of approximate solutions $\{x_t\}_{t=0}^T$ of the problems $(P_T^{(y)})$ in the region containing the right end point of the interval $[0, T]$ was obtained in [52]. It shows that this structure depends neither on x_0 nor T.

Theorem 4.7. *Let M, ϵ be positive numbers and let $L_1 \geq 1$ be an integer. Then there exist a positive number δ and an integer $L_2 > L_1$ such that if an integer $T > 2L_2$ and if an (Ω)-program $\{x_t\}_{t=0}^T$ satisfies*

$$x_0 \in X_M, \ \sum_{t=0}^{T-1} v(x_t, x_{t+1}) \geq \sigma(v, T, x_0) - \delta,$$

then $|x_{T-t} - \Lambda_t(x_*)| \leq \epsilon$ for all integers $t = 0, \ldots, L_1$.

5. Variational Problems with Extended-Valued Integrands

We study turnpike properties of approximate solutions of an autonomous variational problem with a lower semicontinuous integrand $f : R^n \times R^n \to R^1 \cup \{\infty\}$, where R^n is the n-dimensional Euclidean space. More precisely, we consider the following variational problems

$$\int_0^T f(v(t), v'(t))dt \to \min, \tag{P_1}$$

$v : [0, T] \to R^n$ is an absolutely continuous (a.c.) function such that

$$v(0) = x, \ v(T) = y$$

and

$$\int_0^T f(v(t), v'(t))dt \to \min, \tag{P_2}$$

$v : [0, T] \to R^n$ is an a.c. function such that $v(0) = x$,

where $x, y \in R^n$.

We denote by $\mathrm{mes}(E)$ the Lebesgue measure of a Lebesgue measurable set $E \subset R^1$, denote by $|\cdot|$ the Euclidean norm of the space R^n and by $\langle \cdot, \cdot \rangle$

the inner product of R^n. For each function $f : X \to R^1 \cup \{\infty\}$, where X is a nonempty, set

$$\mathrm{dom}(f) = \{x \in X : \ f(x) < \infty\}.$$

Let a be a real positive number, $\psi : [0, \infty) \to [0, \infty)$ be an increasing function such that

$$\lim_{t\to\infty} \psi(t) = \infty$$

and let $f : R^n \times R^n \to R^1 \cup \{\infty\}$ be a lower semicontinuous function such that the set

$$\mathrm{dom}(f) = \{(x, y) \in R^n \times R^n : \ f(x, y) < \infty\}$$

is nonempty, convex and closed and that

$$f(x, y) \geq \max\{\psi(|x|), \ \psi(|y|)|y|\} - a \text{ for each } x, y \in R^n.$$

For each pair of points $x, y \in R^n$ and each positive number T define

$$\sigma(f, T, x) = \inf\{\int_0^T f(v(t), v'(t))dt : \ v : [0, T] \to R^n$$

is an absolutely continuous (a.c.) function satisfying $v(0) = x\}$,

$$\sigma(f, T, x, y) = \inf\{\int_0^T f(v(t), v'(t))dt : \ v : [0, T] \to R^n$$

is an a.c. function satisfying $v(0) = x, \ v(T) = y\}$,

$$\sigma(f, T) = \inf\{\int_0^T f(v(t), v'(t))dt : \ v : [0, T] \to R^n \text{ is an a.c. function}\}.$$

(Here we assume that infimum over an empty set is infinity.)

We suppose that there exists a point $\bar{x} \in R^n$ such that

$$f(\bar{x}, 0) \leq f(x, 0) \text{ for each } x \in R^n \tag{5.1}$$

and that the following assumptions hold:

(A1) $(\bar{x}, 0)$ is an interior point of the set $\mathrm{dom}(f)$ and the function f is continuous at the point $(\bar{x}, 0)$;

(A2) for each positive number M there exists a positive number c_M such that

$$\sigma(f, T, x) \geq Tf(\bar{x}, 0) - c_M$$

for each point $x \in R^n$ satisfying $|x| \leq M$ and each real number $T > 0$;

(A3) for each point $x \in R^n$ the function $f(x, \cdot) : R^n \to R^1 \cup \{\infty\}$ is convex.
Assumption (A2) implies that for each a.c. function $v : [0, \infty) \to R^n$ the function

$$T \to \int_0^T f(v(t), v'(t))dt - Tf(\bar{x}, 0), \ T \in (0, \infty)$$

is bounded from below.

It should be mentioned that inequality (5.1) and assumptions (A1)–(A3) are common in the literature and hold for many infinite horizon optimal control problems [13, 51]. In particular, we need inequality (5.1) and assumption (A2) in the cases when the problems (P_1) and (P_2) possess the turnpike property and the point $\bar{x}$ is its turnpike. Assumption (A2) means that the constant function $\bar{v}(t) = \bar{x}$, $t \in [0, \infty)$ is an approximate solution of the infinite horizon variational problem with the integrand f related to the problems (P_1) and (P_2).

We say that an a.c. function $v : [0, \infty) \to R^n$ is (f)-good [13, 51] if

$$\sup\{|\int_0^T f(v(t), v'(t))dt - Tf(\bar{x}, 0)| : T \in (0, \infty)\} < \infty.$$

The following result was obtained in [53].

Proposition 5.1. *Let $v : [0, \infty) \to R^n$ be an a.c. function. Then either the function v is (f)-good or*

$$\int_0^T f(v(t), v'(t))dt - Tf(\bar{x}, 0) \to \infty \text{ as } T \to \infty.$$

Moreover, if the function v is (f)-good, then $\sup\{|v(t)| : t \in [0, \infty)\} < \infty$.

For each pair of numbers $T_1 \in R^1$, $T_2 > T_1$ and each a.c. function $v : [T_1, T_2] \to R^n$ put

$$I^f(T_1, T_2, v) = \int_{T_1}^{T_2} f(v(t), v'(t))dt.$$

For each positive number M denote by X_M the set of all points $x \in R^n$ such that $|x| \le M$ and there exists an a.c. function $v : [0, \infty) \to R^n$ which satisfies

$$v(0) = x, \ I^f(0, T, v) - Tf(\bar{x}, 0) \le M \text{ for each } T \in (0, \infty).$$

It is clear that $\cup\{X_M : M \in (0, \infty)\}$ is the set of all points $x \in X$ for which there exists an (f)-good function $v : [0, \infty) \to R^n$ such that $v(0) = x$.

We suppose that the following assumption holds:

(A4) (the asymptotic turnpike property) for each (f)-good function $v : [0, \infty) \to R^n$, $\lim_{t\to\infty} |v(t) - \bar{x}| = 0$.

The following turnpike result for the problem (P_2) was established in [53].

Theorem 5.2. *Let ϵ, M be positive numbers. Then there exist an integer $L \geq 1$ and a real number $\delta > 0$ such that for each real number $T > 2L$ and each a.c. function $v : [0, T] \to R^n$ which satisfies*

$$v(0) \in X_M \text{ and } I^f(0, T, v) \leq \sigma(f, T, v(0)) + \delta$$

there exist a pair of numbers $\tau_1 \in [0, L]$ and $\tau_2 \in [T - L, T]$ such that

$$|v(t) - \bar{x}| \leq \epsilon \text{ for all } t \in [\tau_1, \tau_2]$$

and if $|v(0) - \bar{x}| \leq \delta$, then $\tau_1 = 0$.

In the sequel we use a notion of an overtaking optimal function [13, 51].

An a.c. function $v : [0, \infty) \to R^n$ is called (f)-overtaking optimal if for each a.c. function $u : [0, \infty) \to R^n$ satisfying $u(0) = v(0)$ the inequality

$$\limsup_{T\to\infty}[I^f(0, T, v) - I^f(0, T, u)] \leq 0$$

holds.

The following result which establishes the existence of an overtaking optimal function was obtained in [53].

Theorem 5.3. *Assume that $x \in R^n$ and that there exists an (f)-good function $v : [0, \infty) \to R^n$ satisfying $v(0) = x$. Then there exists an (f)-overtaking optimal function $u_* : [0, \infty) \to R^n$ such that $u_*(0) = x$.*

Denote by Card(A) the cardinality of the set A.

Let M be a positive number. Denote by Y_M the set of all points $x \in R^n$ for which there exist a number $T \in (0, M]$ and an a.c. function $v : [0, T] \to R^n$ such that $v(0) = \bar{x}$, $v(T) = x$ and $I^f(0, T, v) \leq M$.

The following turnpike results for the problems (P_1) were established in [58].

Theorem 5.4. *Let $\epsilon, M_0, M_1, M_2 > 0$. Then there exist an integer $Q \geq 1$ and a positive number L such that for each number $T > L$, each point $z_0 \in X_{M_0}$ and each point $z_1 \in Y_{M_1}$, the value $\sigma(f, T, z_0, z_1)$ is finite and for each a.c. function $v : [0, T] \to R^n$ which satisfies*

$$v(0) = z_0, \ v(T) = z_1, \ I^f(0, T, v) \leq \sigma(f, T, z_0, z_1) + M_2$$

there exists a finite sequence of closed intervals $[a_i, b_i] \subset [0, T]$, $i = 1, \dots, q$ *such that*

$$q \le Q, \; b_i - a_i \le L, \; i = 1, \dots, q,$$

$$|v(t) - \bar{x}| \le \epsilon, \; t \in [0, T] \setminus \cup_{i=1}^{q} [a_i, b_i].$$

Theorem 5.5. *Let* ϵ, $M_0, M_1 > 0$. *Then there exist numbers* $L, \; \delta > 0$ *such that for each number* $T > 2L$, *each point* $z_0 \in X_{M_0}$ *and each point* $z_1 \in Y_{M_1}$, *the value* $\sigma(v, T, z_0, z_1)$ *is finite and for each a.c. function* $v : [0, T] \to R^n$ *which satisfies*

$$v(0) = z_0, \; v(T) = z_1, \; I^f(0, T, v) \le \sigma(f, T, z_0, z_1) + \delta$$

there exists a pair of numbers $\tau_1 \in [0, L]$, $\tau_2 \in [T - L, T]$ *such that*

$$|v(t) - \bar{x}| \le \epsilon, \; t \in [\tau_1, \tau_2].$$

Moreover if $|v(0) - \bar{x}| \le \delta$, *then* $\tau_1 = 0$ *and if* $|v(T) - \bar{x}| \le \delta$. *then* $\tau_2 = T$.

6. Two Propositions

In the proofs of the results stated in the previous section the following two propositions play an important role.

Proposition 6.1. *Let* $M_0, M_1 > 0$. *Then there exists a positive number* M_2 *such that for each positive number* T *and each a.c. function* $v : [0, T] \to R^n$ *which satisfies*

$$|v(0)| \le M_0, \; I^f(0, T, v) \le Tf(\bar{x}, 0) + M_1$$

the following inequality holds:

$$|v(t)| \le M_2 \text{ for all } t \in [0, T].$$

Proposition 6.2. *Let* ϵ *be a positive number. Then there exists a positive number* δ *such that if an a.c. function* $v : [0, 1] \to R^n$ *satisfies* $|v(0) - \bar{x}|$, $|v(1) - \bar{x}| \le \delta$, *then*

$$I^f(0, 1, v) \ge f(\bar{x}, 0) - \epsilon.$$

7. Examples

Example 7.1. Let a_0 be a positive number, $\psi_0 : [0, \infty) \to [0, \infty)$ be an increasing function satisfying

$$\lim_{t \to \infty} \psi_0(t) = \infty$$

and let $L : R^n \times R^n \to [0, \infty]$ be a lower semicontinuous function such that

$$\text{dom}(L) := \{(x, y) \in R^n \times R^n : L(x, y) < \infty\}$$

is nonempty, convex, closed set and

$$L(x, y) \geq \max\{\psi_0(|x|), \ \psi_0(|y|)|y|\} - a_0 \text{ for each } x, y \in R^n.$$

Assume that for each point $x \in R^n$ the function $L(x, \cdot) : R^n \to R^1 \cup \{\infty\}$ is convex and that there exists a point $\bar{x} \in R^n$ such that

$$L(x, y) = 0 \text{ if and only if } (x, y) = (\bar{x}, 0),$$

$(\bar{x}, 0)$ is an interior point of $\text{dom}(L)$ and that L is continuous at the point $(\bar{x}, 0)$.

Let $\mu \in R^1$ and $l \in R^n$. Define

$$f(x, y) = L(x, y) + \mu + \langle l, y \rangle, \ \ x, y \in R^n.$$

We showed in [53] that all the assumptions introduced in Sect. 5 hold for f.

Example 7.2. Let a be a positive number, $\psi : [0, \infty) \to [0, \infty)$ be an increasing function such that $\lim_{t \to \infty} \psi(t) = \infty$ and let $f : R^n \times R^n \to R^1 \cup \{\infty\}$ be a convex lower semicontinuous function such that the set $\text{dom}(f)$ is nonempty, convex and closed and that

$$f(x, y) \geq \max\{\psi(|x|), \ \psi(|y|)|y|\} - a \text{ for each } x, y \in R^n.$$

We suppose that there exists a point $\bar{x} \in R^n$ such that

$$f(\bar{x}, 0) \leq f(x, 0) \text{ for each } x \in R^n$$

and that $(\bar{x}, 0)$ is an interior point of the set $\text{dom}(f)$. It is known that the function f is continuous at the point $(\bar{x}, 0)$. It is well-known fact of convex analysis [41] that there exists a point $l \in R^n$ such that

$$f(x, y) \geq f(\bar{x}, 0) + \langle l, y \rangle$$

for each $x, y \in R^n$.

We assume that for each pair of points

$$(x_1, y_1),\ (x_2, y_2) \in \mathrm{dom}(f)$$

satisfying $(x_1, y_1) \neq (x_2, y_2)$ and each number $\alpha \in (0, 1)$, we have

$$f(\alpha(x_1, y_1) + (1 - \alpha)(x_2, y_2)) < \alpha f(x_1, y_1) + (1 - \alpha) f(x_2, y_2).$$

Put

$$L(x, y) = f(x, y) - f(\bar{x}, 0) - \langle l, y\rangle \text{ for each } x, y \in R^n.$$

It is not difficult to see that there exist a positive number a_0 and an increasing function $\psi_0 : [0, \infty) \to [0, \infty)$ such that

$$L(x, y) \geq \max\{\psi_0(|x|),\ \psi_0(|y|)|y|\} - a_0 \text{ for all } x, y \in R^n.$$

It is easy to see that L is a convex, lower semicontinuous function and that the equality $L(x, y) = 0$ holds if and only if $(x, y) = (\bar{x}, 0)$. Now it is easy to see that our example is a particular case of Example 7.1 and all the assumptions introduced in Sect. 5 hold for f.

8. Behavior of Solutions in the Regions Containing End Points

We continue to use the notation and definitions introduced in Sect. 5 and to study the structure of approximate solutions of problems (P_2). Our goal is to study their structure in the regions containing end points.

Let a be a positive number, $\psi : [0, \infty) \to [0, \infty)$ be an increasing function which satisfies

$$\lim_{t \to \infty} \psi(t) = \infty$$

and let $f : R^n \times R^n \to R^1 \cup \{\infty\}$ be a convex lower semicontinuous function such that the set $\mathrm{dom}(f)$ is nonempty and closed and that

$$f(x, y) \geq \max\{\psi(|x|),\ \psi(|y|)|y|\} - a \text{ for each } x, y \in R^n.$$

We suppose that there exists a point $\bar{x} \in R^n$ such that the following assumption holds:

(A5) $(\bar{x}, 0)$ is an interior point of the set $\mathrm{dom}(f)$ and

$$f(\bar{x}, 0) \leq f(x, 0) \text{ for all } x \in R^n.$$

They are well-known facts from convex analysis [41] that the function f is continuous at the point $(\bar{x}, 0)$ and that there exits a point $l \in R^n$ such that

$$f(x, y) \geq f(\bar{x}, 0) + \langle l, y \rangle \text{ for each } x, y \in R^n.$$

We also assume that for each pair of points $(x_1, y_1), \ (x_2, y_2) \in \mathrm{dom}(f)$ such that $(x_1, y_1) \neq (x_2, y_2)$ and each number $\alpha \in (0, 1)$ the inequality

$$f(\alpha(x_1, y_1) + (1 - \alpha)(x_2, y_2)) < \alpha f(x_1, y_1) + (1 - \alpha) f(x_2, y_2)$$

holds. This means that the function f is strictly convex. The integrand f was considered in Example 7.2. It was shown there that assumptions (A1)–(A4) and all the results of Sect. 5 hold for the integrand f.

In our study we use an integrand L defined by

$$L(x, y) = f(x, y) - f(\bar{x}, 0) - \langle l, y \rangle \text{ for all } x, y \in R^n.$$

We suppose that the following assumption holds.

(A6) For each pair of positive numbers M, ϵ there exists a positive number γ such that for each pair of points $(\xi_1, \xi_2), (\eta_1, \eta_2) \in \mathrm{dom}\,(f)$ which satisfies the inequalities $|\xi_i|, |\eta_i| \leq M, i = 1, 2$ and $|\xi_1 - \xi_2| \geq \epsilon$, we have

$$2^{-1} f(\xi_1, \eta_1) + 2^{-1} f(\xi_2, \eta_2) - f(2^{-1}(\xi_1 + \xi_2), 2^{-1}(\eta_1 + \eta_2)) \geq \gamma.$$

Since the restriction of the function f to the set $\mathrm{dom}(f)$ is strictly convex (see assumption (A6)) Theorem 5.3 implies the following result.

Theorem 8.1. *Assume that $x \in R^n$ and that there exists an (f)-good function $v : [0, \infty) \to R^n$ satisfying $v(0) = x$. Then there exists a unique (f)-overtaking optimal function $v_* : [0, \infty) \to R^n$ such that $v_*(0) = x$.*

Let $z \in R^n$ and let there exist an (f)-good function $v : [0, \infty) \to R^n$ such that $v(0) = z$. Denote by $Y^{(f,z)} : [0, \infty) \to R^n$ a unique (f)-overtaking optimal function satisfying $Y^{(f,z)}(0) = z$ which exists by Theorem 8.1.

The following theorem obtained in [54] describes the structure of approximate solutions of variational problems in the regions containing the left end point.

Theorem 8.2. *Let $M, \epsilon > 0$ be real numbers and let $L_0 \geq 1$ be an integer. Then there exist a positive number δ and an integer $L_1 > L_0$ such that for each number $T \geq L_1$, each point $z \in X_M$ and each a.c. function $v : [0, T] \to R^n$ which satisfies*

$$v(0) = z, \ I^f(0, T, v) \leq \sigma(f, T, z) + \delta$$

the inequality

$$|v(t) - Y^{(f,z)}(t)| \leq \epsilon, \ t \in [0, L_0]$$

holds.

We intend to describe the structure of approximate solutions of variational problems in the regions containing the right end point. In order to meet this goal define the functions $\bar{f}, \bar{L} : R^n \times R^n \to R^1 \cup \{\infty\}$ by

$$\bar{f}(x, y) = f(x, -y), \ \bar{L}(x, y) = L(x, -y) \text{ for all } x, y \in R^n.$$

It is not difficult to see that

$$\mathrm{dom}(\bar{f}) = \{(x, y) \in R^n \times R^n : \ (x, -y) \in \mathrm{dom}(f)\},$$

$\mathrm{dom}(\bar{f})$ is nonempty closed convex subset of $R^n \times R^n$,

$$\bar{f}(x, y) \geq \max\{\psi(|x|), \ \psi(|y|)|y|\} - a \text{ for each } x, y \in R^n \times R^n,$$

the point $(\bar{x}, 0)$ is an interior point of the set $\mathrm{dom}(\bar{f})$ and the function $\bar{f}$ is convex and lower semicontinuous.

It is not difficult to see that for each pair of points $x, y \in R^n$,

$$\begin{aligned}\bar{f}(x, y) &= f(x, -y) \geq f(\bar{x}, 0) + \langle l, -y \rangle = \bar{f}(\bar{x}, 0) + \langle -l, y \rangle, \\ \bar{L}(x, y) &= L(x, -y) = f(x, -y) - f(\bar{x}, 0) - \langle l, -y \rangle \\ &= \bar{f}(x, y) - \bar{f}(\bar{x}, 0) - \langle -l, y \rangle\end{aligned}$$

and that for each pair of points $(x_1, y_1), (x_2, y_2) \in \mathrm{dom}(\bar{f})$ such that $(x_1, y_1) \neq (x_2, y_2)$ and each number $\alpha \in (0, 1)$ we have

$$\bar{f}(\alpha(x_1, y_1) + (1 - \alpha)(x_2, y_2)) < \alpha \bar{f}(x_1, y_1) + (1 - \alpha)\bar{f}(x_2, y_2).$$

Therefore all the assumptions posed in this section for the function f also hold for the function $\bar{f}$. Also all the results of this section and of Sect. 5 stated for the function f are valid for the function $\bar{f}$. In particular Theorems 5.2 and 5.3 hold for the integrand $\bar{f}$.

Assumption (A6) imply that the following assumption holds.

(A7) For each pair of numbers $M, \epsilon > 0$ there exists a positive number γ such that for each pair of points $(\xi_1, \xi_2), (\eta_1, \eta_2) \in \mathrm{dom}(\bar{f})$ which satisfies

$$|\xi_i|, \ |\eta_i| \leq M, \ i = 1, 2 \text{ and } |\xi_1 - \xi_2| \geq \epsilon$$

the inequality

$$2^{-1}\bar{f}(\xi_1, \eta_1) + 2^{-1}\bar{f}(\xi_2, \eta_2) - \bar{f}(2^{-1}(\xi_1 + \xi_2), 2^{-1}(\eta_1 + \eta_2)) \geq \gamma_0$$

holds.

It is easy now to see that Theorems 8.1 and 8.2 hold for the integrand $\bar{f}$.

For each positive number M denote by $\bar{X}_M$ the set of all points $x \in R^n$ such that $|x| \leq M$ and that there exists an a.c. function $v : [0, \infty) \to R^n$ which satisfies

$$I^{\bar{f}}(0, T, v) - T\bar{f}(\bar{x}, 0) \leq M \text{ for each } T \in (0, \infty).$$

Set

$$\bar{X}_* = \cup\{\bar{X}_M : M \in (0, \infty)\}.$$

Since the function $\bar{f}$ is convex we conclude that the set $\bar{X}_M$ is convex for all positive numbers M. It is not difficult to show that for each positive number M the set $\bar{X}_M$ is closed.

By Theorem 8.1, applied to the integrand $\bar{f}$, for each point $x \in \bar{X}_*$ there exists a unique $(\bar{f})$-overtaking optimal function $\Lambda^{(x)} : [0, \infty) \to R^n$ such that $\Lambda^{(x)}(0) = x$. Proposition 5.1 implies that $\Lambda^{(x)}$ is $(\bar{f})$-good function for any point $x \in \bar{X}_*$. Therefore, for each point $x \in \bar{X}_*$,

$$\lim_{t \to \infty} |\Lambda^{(x)}(t) - \bar{x}| = 0.$$

For each point $x \in \bar{X}_*$ set

$$\pi(x) = \lim_{T \to \infty} [I^{\bar{f}}(0, T, \Lambda^{(x)}) - T\bar{f}(\bar{x}, 0)].$$

Let $x \in \bar{X}_*$ be given. Then

$$\begin{aligned}\pi(x) &= \lim_{T \to \infty} [\int_0^T \bar{L}(\Lambda^{(x)}(t), (\Lambda^{(x)})'(t))dt - \int_0^T \langle l, (\Lambda^{(x)})'(t)\rangle dt] \\ &= \lim_{T \to \infty} \int_0^T \bar{L}(\Lambda^{(x)}(t), (\Lambda^{(x)})'(t))dt - \lim_{T \to \infty} \langle l, \Lambda^{(x)}(T) - x\rangle \\ &= \int_0^\infty \bar{L}(\Lambda^{(x)}(t), (\Lambda^{(x)})'(t))dt - \langle l, \bar{x} - x\rangle.\end{aligned}$$

Therefore the value $\pi(x)$ is well-defined. Since the function $\Lambda^{(x)}$ is $(\bar{f})$-good, Proposition 5.1 implies that $\pi(x)$ is finite for each $x \in \bar{X}_*$.

The function π plays an important role in our study of the structure of approximate solutions of variational problems in the regions containing the right end point. We show that approximate solutions of the problem (P_2) are arbitrary close to the function $\Lambda^{(x_*)}(T - t)$ in a region which contains the right end point T, where x_* is a unique point of minimum of the function π.

The following result was obtained in [54].

Proposition 8.3. *1. For each positive number M the function $\pi : \bar{X}_M \to R^1$ is lower semicontinuous.*

2. For all pairs of points $y, z \in \bar{X}_*$ satisfying $y \neq z$ and each number $\alpha \in (0, 1)$,

$$\pi(\alpha y + (1 - \alpha)z) < \alpha\pi(y) + (1 - \alpha)\pi(z).$$

3. $\pi(\bar{x}) = 0$.

4. There exists a number $\tilde{M} > |\bar{x}|$ such that $\pi(x) \geq 2$ for each point $x \in \bar{X}_* \setminus \bar{X}_{\tilde{M}}$.

Let a positive number $\tilde{M}$ be as guaranteed by Proposition 8.3. By Proposition 8.3, there exists a unique point $x_* \in \bar{X}_{\tilde{M}}$ such that

$$\pi(x_*) < \pi(x) \text{ for all points } x \in \bar{X}_{\tilde{M}} \setminus \{x_*\}.$$

By Proposition 8.3 if $x \in \bar{X}_* \setminus \bar{X}_{\tilde{M}}$, then

$$\pi(x) \geq 2 > \pi(\bar{x}) \geq \pi(x_*).$$

The following theorem obtained in [54] describes the structure of approximate solutions of variational problems in the regions containing the right end point.

Theorem 8.4. *Let $M, \epsilon > 0$ be real numbers and let $L_1 \geq 1$ be an integer. Then there exist a positive number δ and a natural number $L_2 > L_1$ such that if a number $T > 2L_2$ and if an a.c. function $v : [0, T] \to R^n$ satisfies*

$$v(0) \in X_M \text{ and } I^f(0, T, v) \leq \sigma(f, T, v(0)) + \delta,$$

then

$$|v(T - t) - \Lambda^{(x_*)}(t)| \leq \epsilon \text{ for all } t \in [0, L_1].$$

9. Optimal Solutions for Infinite Horizon Problems

In this section which is based on [56] we study the structure of optimal solutions of infinite horizon autonomous variational problems with a lower semicontinuous integrand $f : R^n \times R^n \to R^1 \cup \{\infty\}$ introduced in Sect. 5. We also show that all the optimality notions used in the literature are equivalent for the problems with the integrand f. We use the notation and definitions introduced in Sect. 5.

Let a be a positive number, $\psi : [0, \infty) \to [0, \infty)$ be an increasing function such that

$$\lim_{t \to \infty} \psi(t) = \infty$$

and let $f : R^n \times R^n \to R^1 \cup \{\infty\}$ be a lower semicontinuous function such that the set dom(f) is nonempty convex and closed and that

$$f(x, y) \geq \max\{\psi(|x|), \ \psi(|y|)|y|\} - a \text{ for each } x, y \in R^n.$$

We suppose that there exists a point $\bar{x} \in R^n$ such that

$$f(\bar{x}, 0) \leq f(x, 0) \text{ for each } x \in R^n$$

and that assumptions (A1)–(A4) introduced in Sect. 5 hold.

We use the notion of an overtaking optimal function introduced in Sect. 5. The following two optimality notions are also used in the infinite horizon optimal control.

An a.c. function $v : [0, \infty) \to R^n$ is called (f)-weakly optimal [11, 51] if for each a.c. function $u : [0, \infty) \to R^n$ satisfying $u(0) = v(0)$, we have

$$\liminf_{T \to \infty} [I^f(0, T, v) - I^f(0, T, u)] \leq 0.$$

An a.c. function $v : [0, \infty) \to R^n$ is called (f)-minimal [5, 51] if for each pair of numbers $T_1 \geq 0$, each $T_2 > T_1$ and each a.c. function $u : [T_1, T_2] \to R^n$ satisfying $u(T_i) = v(T_i)$, $i = 1, 2$, we have

$$\int_{T_1}^{T_2} f(v(t), v'(t))dt \leq \int_{T_1}^{T_2} f(u(t), u'(t))dt.$$

The following theorem obtained in [56] shows that for the integrand considered in the section all the three optimality notions introduced before are equivalent.

Theorem 9.1. *Assume that $x \in R^n$ and that there exists an (f)-good function $\tilde{v} : [0, \infty) \to R^n$ satisfying $\tilde{v}(0) = x$. Let $v : [0, \infty) \to R^n$ be an a.c. function such that $v(0) = x$. Then the following conditions are equivalent:*

(i) the function v is (f)-overtaking optimal; (ii) the function v is (f)-weakly optimal; (iii) the function v is (f)-good and (f)-minimal; (iv) the function v is (f)-minimal and $\lim_{t \to \infty} v(t) = \bar{x}$; (v) the function v is (f)-minimal and $\liminf_{t \to \infty} |v(t) - \bar{x}| = 0$.

The following two theorems obtained in [56] describe the asymptotic behavior of overtaking optimal functions.

Theorem 9.2. *Let ϵ be a positive number. Then there exists a positive number δ such that:*

(i) *For each point $x \in R^n$ satisfying $|x - \bar{x}| \leq \delta$ there exists an (f)-overtaking optimal and (f)-good function $v : [0, \infty) \to R^n$ such that $v(0) = x$.*
(ii) *If an (f)-overtaking optimal function $v : [0, \infty) \to R^n$ satisfies $|v(0) - \bar{x}| \leq \delta$, then $|v(t) - \bar{x}| \leq \epsilon$ for all numbers $t \in [0, \infty)$.*

Theorem 9.3. *Let ϵ, M be positive numbers. Then there exists a number a positive number L such that for each point $x \in X_M$ and each (f)-overtaking optimal function $v : [0, \infty) \to R^n$ satisfying $v(0) = x$ the following inequality holds:*

$$|v(t) - \bar{x}| \leq \epsilon \text{ for all } t \in [L, \infty).$$

The next theorem obtained in [56] establishes a non-self-intersection property of overtaking optimal solutions analogous to the property established in [34, 50] for variational problems with finite-valued integrands.

Theorem 9.4. *Assume that $v : [0, \infty) \to R^n$ is an (f)-good (f)-overtaking optimal function and that $0 \leq t_1 < t_2$ satisfy $v(t_1) = v(t_2)$. Then $v(t) = \bar{x}$ for all numbers $t \geq t_1$.*

10. Agreeable Solutions

We use the notation, definitions and assumptions introduced in Sect. 5. In particular we assume that assumptions (A1)–(A4) hold.

An a.c. function $v_* : [0, \infty) \to R^n$ is called (f)-agreeable [19–21] if for any $T_0 > 0$ and any $\epsilon > 0$ there exists a number $T_\epsilon > T_0$ such that for any number $T \geq T_\epsilon$ there exists an a.c. function $v : [0, T] \to R^n$ which satisfies

$$v(t) = v_*(t), \ t \in [0, T_0],$$

$$I^f(0, T, v) \leq \sigma(f, T, v_*(0)) + \epsilon.$$

We will prove the following result.

Theorem 10.1. *Let $v_* : [0, \infty) \to R^n$ be an a.c. function and let $\tilde{v} : [0, \infty)$ be an (f)-good function such that $v_*(0) = \tilde{v}(0)$. Then the following properties are equivalent:*

(i) the function v_* is (f)-agreeable; (ii) the function v_* is (f)-minimal and

$$\lim_{t \to \infty} |v(t) - \bar{x}| = 0.$$

Theorem 10.1 is proved in the next section.

It is easy to see that Theorems 9.1 and 10.1 imply the following result.

Theorem 10.2. *Assume that* $x \in R^n$ *and that there exists an* (f)*-good function* $\tilde{v} : [0, \infty) \to R^n$ *satisfying* $\tilde{v}(0) = x$. *Let* $v : [0, \infty) \to R^n$ *be an a.c. function such that* $v(0) = x$. *Then the following conditions are equivalent:*

(i) the function v *is* (f)*-overtaking optimal; (ii) the function* v *is* (f)*-weakly optimal; (iii) the function* v *is* (f)*-good and* (f)*-minimal; (iv) the function* v *is* (f)*-minimal and* $\lim_{t\to\infty} v(t) = \bar{x}$; *(v) the function* v *is* (f)*-minimal and* $\liminf_{t\to\infty} |v(t) - \bar{x}| = 0$; *(vi) the function* v *is* (f)*-agreeable.*

11. Proof of Theorem 10.1

Let us show that property (i) implies property (ii). Assume that property (i) holds. We claim that the function v_* is (f)-minimal.

Assume the contrary. Then there exist $S_0 > 0$ and an a.c. function $u : [0, S_0] \to R^n$ such that

$$u(0) = v_*(0), \; u(S_0) = v_*(S_0), \tag{11.1}$$

$$I^f(0, S_0, u) < I^f(0, S_0, v_*).$$

Fix a positive number

$$\Delta < I^f(0, S_0, v_*) - I^f(0, S_0, u). \tag{11.2}$$

Since the function v_* is (f)-agreeable there exists a number $T_1 > S_0$ such that the following property holds:

(P1) for any $T \geq T_1$ there exists an a.c. function $v : [0, T] \to R^n$ such that

$$v(t) = v_*(t), \; t \in [0, S_0], \tag{11.3}$$

$$I^f(0, T, v) \leq \sigma(f, T, v_*(0)) + \Delta/2.$$

By property (P1) there exists an a.c. function $v : [0, T_1] \to R^n$ such that (11.3) holds and

$$I^f(0, T_1, v) \leq \sigma(f, T_1, v_*(0)) + \Delta/2. \tag{11.4}$$

It follows from (11.1) and (11.3) that there exists an a.c. function $z : [0, T_1] \to R^n$ such that

$$z(t) = u(t), \; t \in [0, S_0], \; z(t) = v(t), \; t \in (S_0, T_1]. \tag{11.5}$$

In view of (11.1)–(11.5),

$$
\begin{aligned}
0 &\le I^f(0, T_1, z) - \sigma(f, T_1, v_*(0)) \le I^f(0, T_1, z) - I^f(0, T_1, v) + \Delta/2 \\
&= I^f(0, S_0, u) - I^f(0, S_0, v) + \Delta/2 \\
&< I^f(0, S_0, v_*) - \Delta - I^f(0, S_0, v) + \Delta/2 = -\Delta/2,
\end{aligned}
$$

a contradiction. The contradiction we have reached proves that the function v_* is (f)-minimal.

Let us show that

$$\lim_{t\to\infty} |v_*(t) - \bar{x}| = 0.$$

Let $\epsilon > 0$. Since the function $\tilde{v}$ is (f)-good there exists a constant $M > 0$ such that

$$I^f(0, T, \tilde{v}) \le Tf(\bar{x}, 0) + M \text{ for all } T > 0. \tag{11.6}$$

By (11.6) and the equality $v_*(0) = \tilde{v}(0)$,

$$v_*(0) \in X_M. \tag{11.7}$$

In view of Theorem 5.2, there exist an integer $L \ge 1$ and a number $\delta > 0$ such that the following property holds:

(P2) for each number $T > 2L$ and each a.c. function $v : [0, T] \to R^n$ which satisfies

$$v(0) \in X_M \text{ and } I^f(0, T, v) \le \sigma(f, T, v(0)) + \delta$$

we have

$$|v(t) - \bar{x}| \le \epsilon \text{ for all } t \in [L, T - L].$$

Let a number

$$S > 2L. \tag{11.8}$$

Since v_* is (f)-agreeable there exists $T_\delta > S$ such that the following property holds:

(P3) for any $T \ge T_\delta$ there exists an a.c. function $v : [0, T] \to R^n$ which satisfies

$$v(t) = v_*(t),\ t \in [0, S], \tag{11.9}$$

$$I^f(0, T, v) \le \sigma(f, T, v_*(0)) + \delta. \tag{11.10}$$

Let $T \ge T_\delta$ and let an a.c. function $v : [0, T] \to R^n$ be as guaranteed by (P3). Thus (11.9) and (11.10) hold.

It follows from (11.7)–(11.10) and property (P2),

$$|v(t) - \bar{x}| \le \epsilon \text{ for all } t \in [L, T - L].$$

Together with (11.8) and (11.9) this implies that

$$|v_*(t) - \bar{x}| \le \epsilon \text{ for all } t \in [L, S - L].$$

Since the inequality above holds for any $S > 2L$ we have

$$|v_*(t) - \bar{x}| \le \epsilon \text{ for all } t \ge L.$$

Thus

$$\lim_{t \to \infty} |v_*(t) - \bar{x}| = 0$$

and we showed that property (ii) holds.

Assume that the function v_* is (f)-minimal and that

$$\lim_{t \to \infty} |v_*(t) - \bar{x}| = 0.$$

We claim that the function v_* is (f)-agreeable. By Theorem 9.1 the function v_* is (f)-good and there exists a number $M > 0$ such that

$$|v_*(0)| < M, \tag{11.11}$$

$$|I^f(0, T, v_*) - Tf(\bar{x}, 0)| < M \text{ for all } T > 0. \tag{11.12}$$

Let $T_0 > 0$ and $\epsilon > 0$. By Proposition 6.2 there exists a positive number $\epsilon_0 < \epsilon$ such that the following property holds:

(P4) for each function $v : [0, 1] \to R^n$ satisfying

$$|v(0) - \bar{x}|, \ |v(1) - \bar{x}| \le \epsilon_0,$$

we have

$$I^f(0, 1, v) \ge f(\bar{x}, 0) - \epsilon/8.$$

In view of (A1) there exists a number $\epsilon_1 \in (0, \epsilon_0)$ such that the following property holds:

(P5)

$$\{(z, y) \in R^n \times R^n : |z - \bar{x}| \le 4\epsilon_1, \ |y| \le 4\epsilon_1\} \subset \text{dom}(f),$$

$$|f(z, y) - f(\bar{x}, 0)| \le \epsilon/8 \text{ for each } (y, z) \in R^n \times R^n$$

$$\text{satisfying } |\bar{x} - z| \le 4\epsilon_1, \ |y| \le 4\epsilon_1.$$

In view of (A4) there exists $L_0 > 0$ such that

$$|v_*(t) - \bar{x}| \le \epsilon_1 \text{ for all } t \ge L_0. \tag{11.13}$$

By Theorem 5.2 there exist an integer $L_1 \geq 1$ and a number $\delta \in (0, \epsilon/4)$ such that the following property holds:
(P6) for each number $T > 2L_1$ and each a.c. function $v : [0, T] \to R^n$ which satisfies

$$v(0) \in X_M \text{ and } I^f(0, T, v) \leq \sigma(f, T, v(0)) + \delta$$

we have

$$|v(t) - \bar{x}| \leq \epsilon_1 \text{ for all } t \in [L_1, T - L_1].$$

Fix

$$T_\epsilon > 2L_1 + 2L_0 + T_0 + 4. \tag{11.14}$$

Let a number $T \geq T_\epsilon$. By (11.12) and (11.14), there exists an a.c. function $v : [0, T] \to R^n$ which satisfies

$$v(0) = v_*(0), \tag{11.15}$$
$$I^f(0, T, v) \leq \sigma(f, T, v_*(0)) + \delta. \tag{11.16}$$

It follows from (11.12), (11.14), (11.15), (11.16) and (P6) that

$$|v(t) - \bar{x}| \leq \epsilon_1 \text{ for all } t \in [L_1, T - L_1]. \tag{11.17}$$

By (11.14), (11.17) and (P4),

$$I^f(T - L_1 - 2, T - L_1 - 1, v) \geq f(\bar{x}, 0) - \epsilon/8. \tag{11.18}$$

Define an a.c. function $\hat{v} : [0, T - L_1 - 1] \to R^n$ by

$$\hat{v}(t) = v(t), \; t \in [0, T - L_1 - 2],$$

$$\hat{v}(t) = v(T - L_1 - 2) + (t - (T - L_1 - 2))[v_*(T - L_1 - 1) - v(T - L_1 - 2)], \\ t \in (T - L_1 - 2, T - L_1 - 1]. \tag{11.19}$$

By (11.15) and (11.19),

$$\hat{v}(0) = v_*(0), \; \hat{v}(T - L_1 - 1) = v_*(T - L_1 - 1). \tag{11.20}$$

Since the function v_* is (f)-minimal (11.20) implies that

$$I^f(0, T - L_1 - 1, v_*) \leq I^f(0, T - L_1 - 1, \hat{v}). \tag{11.21}$$

It follows from (11.13), (11.14), (11.17) and (11.19) that for all $t \in (T - L_1 - 2, T - L_1 - 1)$,

$$\hat{v}'(t) = v_*(T - L_1 - 1) - v(T - L_1 - 2), \tag{11.22}$$

$$|\hat{v}'(t)| \le |v_*(T - L_1 - 1) - \bar{x}| + |\bar{x} - v(T - L_1 - 2)| \le 2\epsilon_1, \tag{11.23}$$

$$|\hat{v}(t) - \bar{x}| \le |v(T - L_1 - 2) - \bar{x}| + |\hat{v}'(t)| \le 3\epsilon_1. \tag{11.24}$$

In view of (11.23), (11.24) and (P5), for all $t \in (T - L_1 - 2, T - L_1 - 1)$,

$$|f(\hat{v}(t), \hat{v}'(t)) - f(\bar{x}, 0)| \le \epsilon/8$$

and

$$I^f(T - L_1 - 2, T - L_1 - 1, \hat{v}) \le f(\bar{x}, 0) + \epsilon/8. \tag{11.25}$$

In view of (11.19), (11.21) and (11.25),

$$\begin{aligned} I^f(0, T - L_1 - 1, v_*) &\le I^f(0, T - L_1 - 2, v) + f(\bar{x}, 0) + \epsilon/8 \\ &= I^f(0, T - L_1 - 1, v) - I^f(T - L_1 - 2, T - L_1 - 1, v) + f(\bar{x}, 0) + \epsilon/8 \\ &\le I^f(0, T - L_1 - 1, v) + \epsilon/4. \end{aligned} \tag{11.26}$$

By (11.14), (11.17) and (P4),

$$I^f(T - L_1 - 1, T - L_1, v) \ge f(\bar{x}, 0) - \epsilon/8. \tag{11.27}$$

Define an a.c. function $u : [0, T - L_1] \to R^n$ by

$$u(t) = v_*(t), \ t \in [0, T - L_1 - 1],$$

$$u(t) = v_*(T - L_1 - 1) + (t - (T - L_1 - 1))[v(T - L_1) - v_*(T - L_1 - 1)], \quad t \in (T - L_1 - 1, T - L_1]. \tag{11.28}$$

By (11.28) and (11.15),

$$u(0) = v(0), \ u(T - L_1) = v(T - L_1). \tag{11.29}$$

Set

$$u(t) = v(t), \ t \in (T - L_1, T]. \tag{11.30}$$

It follows from (11.13), (11.14), (11.17) and (11.28) that for all $t \in (T - L_1 - 1, T - L_1)$,

$$u'(t) = v(T - L_1) - v_*(T - L_1 - 1),$$

$$|u'(t)| \le |v(T - L_1) - \bar{x}| + |\bar{x} - v_*(T - L_1 - 1)] \le 2\epsilon_1, \tag{11.31}$$

$$|u(t) - \bar{x}| \le |v_*(T - L_1 - 1) - \bar{x}| + |v(T - L_1) - v_*(T - L_1 - 1)| \le 3\epsilon_1. \tag{11.32}$$

In view of (11.31), (11.32) and (P5), for all $t \in (T - L_1 - 1, T - L_1)$,

$$|f(u(t), u'(t)) - f(\bar{x}, 0)| \le \epsilon/8$$

and

$$I^f(T - L_1 - 1, T - L_1, u) \le f(\bar{x}, 0) + \epsilon/8. \tag{11.33}$$

In view of (11.16), (11.26), (11.27), (11.28), (11.30) and (11.33),

$$\begin{aligned}
I^f(0, T, u) &= I^f(0, T - L_1, u) + I^f(T - L_1, T, v)\\
&\le I^f(0, T - L_1 - 1, v_*) + f(\bar{x}, 0) + \epsilon/8 + I^f(T - L_1, T, v)\\
&\le I^f(0, T - L_1 - 1, v) + \epsilon/4 + I^f(T - L_1 - 1, T - L_1, v)\\
&\quad +\epsilon/4 + I^f(T - L_1, T, v)\\
&= I^f(0, T, v) + \epsilon/2 \le \sigma(f, T, v_*(0)) + \epsilon/4/ + \epsilon/2
\end{aligned}$$

and

$$I^f(0, T, u) \le \sigma(f, T, v_*(0)) + \epsilon. \tag{11.34}$$

Thus in view of (11.28) for any $T \ge T_\epsilon$ there exists an a.c. function $u : [0, T] \to R^n$ such that $u(t) = v_*(t),\ t \in [0, T_0]$ and (11.34) holds. Thus the function v_* is (f)-agreeable. This completes the proof of Theorem 10.1.

References

1. Anderson, B.D.O., Moore, J.B.: Linear Optimal Control. Prentice-Hall, Englewood Cliffs (1971)
2. Aseev, S.M., Kryazhimskiy, A.V.: The Pontryagin Maximum principle and transversality conditions for a class of optimal control problems with infinite time horizons. SIAM J. Control Optim. **43**, 1094–1119 (2004)

3. Aseev, S.M., Veliov, V.M.: Maximum principle for infinite-horizon optimal control problems with dominating discount. Dyn. Contin. Discrete Impulsive Syst. Ser. B **19**, 43–63 (2012)
4. Atsumi, H.: Neoclassical growth and the efficient program of capital accumulation. Rev. Econ. Stud. **32**, 127–136 (1965)
5. Aubry, S., Le Daeron, P.Y.: The discrete Frenkel-Kontorova model and its extensions I. Physica D **8**, 381–422 (1983)
6. Baumeister, J., Leitao, A., Silva, G.N.: On the value function for nonautonomous optimal control problem with infinite horizon. Syst. Control Lett. **56**, 188–196 (2007)
7. Blot, J.: Infinite-horizon Pontryagin principles without invertibility. J. Nonlinear Convex Anal. **10**, 177–189 (2009)
8. Blot, J., Cartigny, P.: Optimality in infinite-horizon variational problems under sign conditions. J. Optim. Theory Appl. **106**, 411–419 (2000)
9. Blot, J., Hayek, N.: Sufficient conditions for infinite-horizon calculus of variations problems. ESAIM Control Optim. Calc. Var. **5**, 279–292 (2000)
10. Bright, I.: A reduction of topological infinite-horizon optimization to periodic optimization in a class of compact 2-manifolds. J. Math. Anal. Appl. **394**, 84–101 (2012)
11. Brock, W.A.: On existence of weakly maximal programmes in a multi-sector economy. Rev. Econ. Stud. **37**, 275–280 (1970)
12. Carlson, D.A.: The existence of catching-up optimal solutions for a class of infinite horizon optimal control problems with time delay. SIAM J. Control Optim. **28**, 402–422 (1990)
13. Carlson, D.A., Haurie, A., Leizarowitz, A.: Infinite Horizon Optimal Control. Springer, Berlin (1991)
14. Cartigny, P., Michel, P.: On a sufficient transversality condition for infinite horizon optimal control problems. Automatica **39**, 1007–1010 (2003)
15. Coleman, B.D., Marcus, M., Mizel, V.J.: On the thermodynamics of periodic phases. Arch. Ration. Mech. Anal. **117**, 321–347 (1992)
16. Gaitsgory, V., Rossomakhine, S., Thatcher, N.: Approximate solution of the HJB inequality related to the infinite horizon optimal control problem with discounting. Dyn. Contin. Discrete Impulsive Syst. Ser. B **19**, 65–92 (2012)
17. Gale, D.: On optimal development in a multi-sector economy. Rev. Econ. Stud. **34**, 1–18 (1967)
18. Guo, X., Hernandez-Lerma, O.: Zero-sum continuous-time Markov games with unbounded transition and discounted payoff rates. Bernoulli **11**, 1009–1029 (2005)

19. Hammond, P.J.: Consistent planning and intertemporal welfare economics. University of Cambridge, Cambridge (1974)
20. Hammond, P.J.: Agreeable plans with many capital goods. Rev. Econ. Stud. **42**, 1–14 (1975)
21. Hammond, P.J., Mirrlees, J.A.: Agreeable plans. In: Models of Economic Growth, pp. 283–299. Wiley, New York (1973)
22. Hayek, N.: Infinite horizon multiobjective optimal control problems in the discrete time case. Optimization **60**, 509–529 (2011)
23. Jasso-Fuentes, H., Hernandez-Lerma, O.: Characterizations of overtaking optimality for controlled diffusion processes. Appl. Math. Optim. **57**, 349–369 (2008)
24. Khan, M.A., Zaslavski, A.J.: On two classical turnpike results for the Robinson-Solow-Srinivasan (RSS) model. Adv. Math. Econ. **13**, 47–97 (2010)
25. Kolokoltsov, V., Yang, W.: The turnpike theorems for Markov games. Dyn. Games Appl. **2**, 294–312 (2012)
26. Leizarowitz, A.: Infinite horizon autonomous systems with unbounded cost. Appl. Math. Opt. **13**, 19–43 (1985)
27. Leizarowitz, A.: Tracking nonperiodic trajectories with the overtaking criterion. Appl. Math. Opt. **14**, 155–171 (1986)
28. Leizarowitz, A., Mizel, V.J.: One dimensional infinite horizon variational problems arising in continuum mechanics. Arch. Ration. Mech. Anal. **106**, 161–194 (1989)
29. Lykina, V., Pickenhain, S., Wagner, M.: Different interpretations of the improper integral objective in an infinite horizon control problem. J. Math. Anal. Appl. **340**, 498–510 (2008)
30. Makarov, V.L., Rubinov, A.M.: Mathematical Theory of Economic Dynamics and Equilibria. Springer, New York (1977)
31. Malinowska, A.B., Martins, N., Torres, D.F.M.: Transversality conditions for infinite horizon variational problems on time scales. Optim. Lett. **5**, 41–53 (2011)
32. Marcus, M., Zaslavski, A.J.: On a class of second order variational problems with constraints. Isr. J. Math. **111**, 1–28 (1999)
33. Marcus, M., Zaslavski, A.J.: The structure of extremals of a class of second order variational problems. Ann. Inst. H. Poincaré Anal. Nonlinéaire **16**, 593–629 (1999)
34. Marcus, M., Zaslavski, A.J.: The structure and limiting behavior of locally optimal minimizers. Ann. Inst. H. Poincaré Anal. Nonlinéaire **19**, 343–370 (2002)
35. McKenzie, L.W.: Turnpike theory. Econometrica **44**, 841–866 (1976)
36. Mordukhovich, B.S.: Minimax design for a class of distributed parameter systems. Automat. Remote Control **50**, 1333–1340 (1990)

37. Mordukhovich, B.S.: Optimal control and feedback design of state-constrained parabolic systems in uncertainly conditions. Appl. Anal **90**, 1075–1109 (2011)
38. Mordukhovich, B.S., Shvartsman, I.: Optimization and feedback control of constrained parabolic systems under uncertain perturbations. In: Optimal Control, Stabilization and Nonsmooth Analysis. Lecture Notes in Control and Information Science, pp. 121–132. Springer, Berlin (2004)
39. Ocana Anaya, E., Cartigny, P., Loisel, P.: Singular infinite horizon calculus of variations. Applications to fisheries management. J. Nonlinear Convex Anal. **10**, 157–176 (2009)
40. Pickenhain, S., Lykina, V., Wagner, M.: On the lower semicontinuity of functionals involving Lebesgue or improper Riemann integrals in infinite horizon optimal control problems. Control Cybern. **37**, 451–468 (2008)
41. Rockafellar, R.T.: Convex Analysis. Princeton University Press, Princeton (1970)
42. Rubinov, A.M.: Economic dynamics. J. Sov. Math. **26**, 1975–2012 (1984)
43. Samuelson, P.A.: A catenary turnpike theorem involving consumption and the golden rule. Am. Econ. Rev. **55**, 486–496 (1965)
44. von Weizsacker, C.C.: Existence of optimal programs of accumulation for an infinite horizon. Rev. Econ. Stud. **32**, 85–104 (1965)
45. Zaslavski, A.J.: Ground states in Frenkel-Kontorova model. Math. USSR Izv. **29**, 323–354 (1987)
46. Zaslavski, A.J.: Optimal programs on infinite horizon 1. SIAM J. Control Optim. **33**, 1643–1660 (1995)
47. Zaslavski, A.J.: Optimal programs on infinite horizon 2. SIAM J. Control Optim. **33**, 1661–1686 (1995)
48. Zaslavski, A.J.: Turnpike theorem for convex infinite dimensional discrete-time control systems. Convex Anal. **5**, 237–248 (1998)
49. Zaslavski, A.J.: Turnpike theorem for nonautonomous infinite dimensional discrete-time control systems. Optimization **48**, 69–92 (2000)
50. Zaslavski, A.J.: A nonintersection property for extremals of variational problems with vector-valued functions. Ann. Inst. H. Poincare Anal. Nonlineare **23**, 929–948 (2006)
51. Zaslavski, A.J.: Turnpike Properties in the Calculus of Variations and Optimal Control. Springer, New York (2006)
52. Zaslavski, A.J.: Turnpike results for a discrete-time optimal control systems arising in economic dynamics. Nonlinear Anal. **67**, 2024–2049 (2007)
53. Zaslavski, A.J.: A turnpike result for a class of problems of the calculus of variations with extended-valued integrands. J. Convex Anal. **15**, 869–890 (2008)

54. Zaslavski, A.J.: Structure of approximate solutions of variational problems with extended-valued convex integrands. ESAIM Control Optim. Calc. Var. **15**, 872–894 (2009)
55. Zaslavski, A.J.: Two turnpike results for a discrete-time optimal control systems. Nonlinear Anal. **71**, 902–909 (2009)
56. Zaslavski, A.J.: Optimal solutions for a class of infinite horizon variational problems with extended-valued integrands. Optimization **59**, 181–197 (2010)
57. Zaslavski, A.J.: Turnpike properties of approximate solutions for discrete-time control systems. Commun. Math. Anal. **11**, 36–45 (2011)
58. Zaslavski, A.J.: Two turnpike results for a continuous-time optimal control systems. In: Proceedings of an International Conference, Complex Analysis and Dynamical Systems IV: Function Theory and Optimization, vol. 553, pp. 305–317 (2011)
59. Zaslavski, A.J.: Structure of Solutions of Variational Problems. SpringerBriefs in Optimization. Springer, New York (2013)
60. Zaslavski, A.J., Leizarowitz, A.: Optimal solutions of linear control systems with nonperiodic integrands. Math. Oper. Res. **22**, 726–746 (1997)

Adv. Math. Econ. 18, 135–140 (2014)

Advances in MATHEMATICAL ECONOMICS

A Characterization of Quasi-concave Function in View of the Integrability Theory

Yuhki Hosoya

Graduate School of Economics, Keio University,
2-15-45 Mita, Minato-ku, Tokyo 108-8345, Japan
(e-mail: ukki@gs.econ.keio.ac.jp)

Received: February 26, 2013
Revised: December 2, 2013

JEL classification: D11

Mathematics Subject Classification (2010): 91B08, 91B16

Abstract. Let g and u be C^1-class real-valued functions that satisfy the Lagrange multiplier condition $Du = \lambda g$ and $Du \neq 0$. In this paper, we show that u is quasi-concave if and only if g satisfies an inequality which is related to the Bordered Hessian condition even if both of u and g are C^1 rather than C^2.

Key words: Integrability, Inverse demand function, Quasi-concavity, Utility function

1. Introduction

Let C be an open and convex subset of $\mathbb{R}^n$ and $g : C \to \mathbb{R}^n \setminus \{0\}$ be a C^1-function. Hosoya [2] shows that under integrability condition of g, for any $x \in C$ there exist an open (and convex, if necessary) neighborhood U of x and a couple of functions $u : U \to \mathbb{R}$ and $\lambda : U \to \mathbb{R}$ such that

(a) u is C^1-class,
(b) λ is positive and continuous, and furthermore,
(c) $Du(y) = \lambda(y)g(y)$ for any $y \in U$.

However, we have to admit that it is an incomplete result for the theory of consumers in the following sense. Usually, g and u are interpreted as an inverse demand function and a utility function, respectively. Hence, it is required that every $y \in U$ is a maximum point of u with the budget constraint determined by the price $g(y)$ and the income $g(y) \cdot y$. That is, each $y \in U$ must be a solution of the following problem:

$$\begin{aligned} \max \quad & u(z) \\ \text{subject to.} \quad & g(y) \cdot z \le g(y) \cdot y, \\ & z \in U. \end{aligned} \tag{1}$$

We say that u has the **property A** if it satisfies the above condition.

In this paper, we assume that two functions $u, \lambda : U \to \mathbb{R}$ with conditions (a), (b) and (c) exist, where U is an open and convex subset U of C. Our main task is to show the following results: (1) u satisfies the property A if and only if u is quasi-concave,[1] and (2) u is quasi-concave if and only if

$$w^T Dg(y)w \le 0$$

for any $y \in U$ and any $w \in \mathbb{R}^n$ such that $w \cdot g(y) = 0$.

Note that if u is C^2, the result of (2) is an easy consequence of the result of Otani [3]. Indeed, Otani [3] shows that u is quasi-concave if and only if

$$w^T D^2u(y)w \le 0$$

for any $y \in U$ and any $w \in \mathbb{R}^n$ such that $w \cdot Du(y) = 0$.[2] Under conditions (a), (b) and (c), $w \cdot Du(y) = 0$ if and only if $w \cdot g(y) = 0$. Also, under conditions (a), (b) and (c), λ is C^1 if u is C^2 and thus,

$$\begin{aligned} w^T D^2u(y)w &= w^T[g(y)D\lambda(y) + \lambda(y)Dg(y)]w \\ &= \lambda(y)w^T Dg(y)w \end{aligned}$$

for any $w \in \mathbb{R}^n$ such that $w \cdot g(y) = 0$. Hence, (2) holds in this case. Therefore, our main purpose is to extend this result for the case u is C^1 rather than C^2, provided that the integrability condition is fulfilled.[3]

In Sect. 2, we present a rigorous statement of our theorem. The proof of theorem is in Sect. 3.

[1] A real-valued function f defined on the convex set is said to be **quasi-concave** if every upper contour set $\{x | f(x) \ge a\}$ is convex.

[2] This condition is known as the Bordered Hessian condition.

[3] Debreu [1] introduces an example of C^1-class function g such that it satisfies the integrability condition and there is no C^2-function u such that $Du = \lambda g$ for some C^1-function λ. Hence, our extension is meaningful.

2. Main Results

Let $n \geq 2$, $U \subset \mathbb{R}^n$ be open and convex, and the functions $g : U \to \mathbb{R}^n \setminus \{0\}$, $\lambda : U \to \mathbb{R}_{++}$ and $u : U \to \mathbb{R}$ be given. We assume both g and u are C^1 and,

$$Du(x) = \lambda(x)g(x),$$

for any $x \in U$.

Theorem. *The following three claims are equivalent.*

(1) u has property A.
(2) u is quasi-concave.
(3) $w^T Dg(x)w \leq 0$ for any $x \in U$ and $w \in \mathbb{R}^n$ such that $w \cdot g(x) = 0$.

3. Proof

3.1. (2) Implies (1)

Suppose that u is quasi-concave and fix $x \in U$. Then, $Du(x) \neq 0$ and $Du(x) = \lambda(x)g(x)$, that is, the Lagrange multiplier condition holds at x. Hence, u has property A and thus (2) implies (1). ■

3.2. (1) Implies (3)

Suppose (1) holds and $w \cdot g(x) = 0$. Define $x(t) = x + tw$. Then,

$$g(x) \cdot x(t) = g(x) \cdot x$$

for any t. Let $c(t) = u(x(t))$. Since (1) holds and U is open, the function c can be defined on $[-\varepsilon, \varepsilon]$ for sufficiently small $\varepsilon > 0$, and

$$c(t) \leq c(0)$$

for any $t \in [-\varepsilon, \varepsilon]$. Note that $c'(t) = Du(x(t)) \cdot w$. By the mean value theorem, there exists a sequence (t_m) such that $t_m \downarrow 0$ and $c'(t_m) \leq 0$. Hence, we have the following evaluation:

$$\begin{aligned}
0 &\geq \limsup_{m\to\infty} \frac{Du(x(t_m)) \cdot w}{t_m} \\
&= \limsup_{m\to\infty} \frac{\lambda(x(t_m))g(x(t_m)) \cdot w}{t_m} \\
&\geq M \limsup_{m\to\infty} \frac{g(x(t_m)) \cdot w}{t_m}
\end{aligned}$$

$$= M \limsup_{m\to\infty} \frac{g(x(t_m)) - g(x(0))}{t_m} \cdot w$$
$$= M w^T Dg(x) w,$$

where $M = \max_{t\in[-\varepsilon,\varepsilon]} \lambda(x(t)) > 0$. Consequently, we have $w^T Dg(x)w \leq 0$, and thus, (3) holds. ■

3.3. (3) Implies (2)

Suppose (3) holds and (2) does not hold. Then, there exists $x, y \in U$ and $z \in [x, y]$ such that $u(z) < \min\{u(x), u(y)\}$. Define a number s^* by

$$s^* = \max[\arg\min_{t\in[0,1]} u((1-t)x + ty)].$$

By the assumption concerning x and y, we have $0 \neq s^* \neq 1$. Let $x(t) = (1-t)x + ty$, $w = x(s^*)$ and $Du(w) = p$. Then $p \neq 0$ and thus $\|p\| \neq 0$.

Consider the following function:

$$f(a, b) = u(w + a(y - x) + bp) = u(x(s^* + a) + bp).$$

Then, f is C^1 around $(0, 0)$, $f(0, 0) = u(w)$ and $f_b(0, 0) = \|p\|^2 > 0$, and thus, by the implicit function theorem, there exists $\varepsilon_1 > 0$, $\varepsilon_2 > 0$ and a C^1 function $b : [-\varepsilon_1, \varepsilon_1] \to \mathbb{R}$ such that $b(0) = 0$ and $f(a, b) = u(w)$ if and only if $b = b(a)$ for any $(a, b) \in [-\varepsilon_1, \varepsilon_1] \times [-\varepsilon_2, \varepsilon_2]$. Now, since

$$b'(a) = -\frac{f_a(a, b(a))}{f_b(a, b(a))}$$
$$= -\frac{Du(x(s^* + a) + b(a)p) \cdot (y - x)}{Du(x(s^* + a) + b(a)p) \cdot p},$$

we have

$$b'(0) = -\frac{Du(w) \cdot (y - x)}{Du(w) \cdot p} = 0$$

by the first-order condition of the following minimization problem:

$$\min_{t\in[0,1]} u(x(t)).$$

Meanwhile, since $f(a, b(a)) = u(w)$, we have

$$Du(x(s^* + a) + b(a)p) \cdot [b'(a)p + (y - x)] = 0,$$

and thus,

$$g(x(s^* + a) + b(a)p) \cdot [b'(a)p + (y - x)] = 0$$

for any $a \in [-\varepsilon_1, \varepsilon_1]$.

Clearly, if $a = 0$, then $Du(x(s^* + a) + b(a)p) \cdot p = \|p\|^2 > 0$. Hence, if $a > 0$ is sufficiently small, then $Du(x(s^* + a) + b(a)p) \cdot p > 0$. For such $a > 0$,

$$\begin{aligned}
0 = \limsup_{a' \downarrow a} \frac{1}{a' - a} &\{Du(x(s^* + a) + b(a)p) \cdot [b'(a')p + (y - x)] \\
&-Du(x(s^* + a) + b(a)p) \cdot [b'(a')p + (y - x)]\} \\
= \limsup_{a' \downarrow a} \frac{1}{a' - a} &\{Du(x(s^* + a) + b(a)p) \cdot [b'(a')p + (y - x)] \\
&-\lambda(x(s^* + a) + b(a)p)g(x(s^* + a) + b(a)p) \cdot [b'(a')p + (y - x)]\} \\
= \limsup_{a' \downarrow a} \frac{1}{a' - a} &\{Du(x(s^* + a) + b(a)p) \cdot [(b'(a')p + (y - x)) \\
&-(b'(a)p + (y - x))] + \lambda(x(s^* + a) + b(a)p)[g(x(s^* + a') + b(a')p) \\
&-g(x(s^* + a) + b(a)p)] \cdot [b'(a')p + (y - x)]\} \\
= Du(x(s^* + a) + b(a)) \cdot p \times \limsup_{a' \downarrow a} &\frac{b'(a') - b'(a)}{a' - a} \\
&+\lambda(x(s^* + a) + b(a)p)[b'(a)p + (y - x)]^T Dg(x(s^* + a) + b(a)p) \\
&[b'(a)p + (y - x)],
\end{aligned}$$

where the second term of the right-hand side is non-positive from (3). Hence, we have

$$\limsup_{a' \downarrow a} \frac{b'(a') - b'(a)}{a' - a} \ge 0.$$

Similarly, we can show that

$$\limsup_{a' \uparrow a} \frac{b'(a') - b'(a)}{a' - a} \ge 0$$

for sufficiently small $a > 0$.

Now, fix any such $a > 0$ and define $h(\theta) = b'(\theta)a - b'(a)\theta$. Then $h(a) = h(0) = 0$, and thus, there exists $\theta^* \in]0, a[$ such that either $h(\theta) \le h(\theta^*)$ for any $\theta \in [0, a]$ or $h(\theta) \ge h(\theta^*)$ for any $\theta \in [0, a]$. If the former holds, then

$$\begin{aligned}
0 &\ge \limsup_{\theta \downarrow \theta^*} \frac{h(\theta) - h(\theta^*)}{\theta - \theta^*} \\
&= a \limsup_{\theta \downarrow \theta^*} \frac{b'(\theta) - b'(\theta^*)}{\theta - \theta^*} - b'(a),
\end{aligned}$$

and thus, we have $b'(a) \geq 0$. We can show $b'(a) \geq 0$ in the latter case similarly. Therefore, we have $b'(a) \geq 0$ for any sufficiently small $a > 0$. Since $b(0) = 0$, we have $b(a) \geq 0$ for any sufficiently small $a > 0$.

Now, since

$$f_b(0,0) = \|p\|^2 > 0,$$

there exists a neighborhood V of $(0,0)$ such that $f_b(a,b) > 0$ for any $(a,b) \in V$. If $a > 0$ is sufficiently small, then $(a,b) \in V$ for any $b \in [0, b(a)]$. Therefore,

$$Du(x(s^* + a) + bp) \cdot p = f_b(a,b) > 0,$$

and thus,

$$u(x(s^* + a)) \leq u(x(s^* + a) + b(a)p) = u(w),$$

which contradicts the definition of s^*. Hence, we conclude that (3) implies (2). ■

Acknowledgements We are grateful to Shinichi Suda and Toru Maruyama for their helpful comments and suggestions.

References

1. Debreu, G.: Smooth preferences, a corrigendum. Econometrica **44**, 831–832 (1976)
2. Hosoya, Y.: Elementary form and proof of the Frobenius theorem for economists. Adv. Math. Econ. **16**, 39–52 (2012)
3. Otani, K.: A characterization of quasi-concave functions. J. Econ. Theory **31**, 194–196 (1983)

Adv. Math. Econ. 18, 141–142 (2014)

Advances in
MATHEMATICAL
ECONOMICS

Subject Index

A
American option, 81
Asymptotic turnpike property, 102, 105, 115

B
Bermuda derivatives, 93
Bermuda type, 89–93
Boundedness, 23
Bounded variation (BVC), 58

C
Caratheodory, 31
Cardinality, 107
Closure type lemma, 38
Compact metric space, 107
Control measures, 18

D
Decomposable, 24
Dependence, 35
Differential game, 1
Discrete-time optimal control system, 101
Distance functions, 53
Dynamic programming principle (DPP), 20
DPP property, 50

E
Eberlein–Smulian theorem, 14
Equicontinuous, 15
Euclidean space, 108
Evolution inclusion, 18
Extreme points, 13

F
(f)-agreeable, 124, 127, 130
(f)-good function, 119
(f)-overtaking optimal function, 115, 119
First-order condition, 138

G
Green function, 1
Grothendieck, 29

H
HJB equation, 53
Hörmander type diffusion processes, 62

I
Implicit function theorem, 138
Increasing function, 117
Infinite horizon optimal control problems, 114
Integrability condition, 135
Integral functionals, 18
Integral representation, 12
Integrands, 40, 120

S. Kusuoka and T. Maruyama (eds.), *Advances in Mathematical Economics Volume 18*, DOI: 10.1007/978-4-431-54834-8,

L
Lagrange multiplier condition, 137
Lebesgue dominated convergence, 37
Lebesgue measure, 112
Local maximum, 43, 56
Local minimum, 49
Lower semicontinuity, 18
Lower semicontinuous, 112
Lower semicontinuous function, 118
Lower semicontinuous integrands, 112
Lower–upper value function, 44
Lyapunov theorem, 13

M
Mackey topology, 29
Minimal, 123, 126, 128
Modelisation, 52
m-point boundary, 28
m-point boundary problem, 12
Multifunction, 38

N
Narrow topology, 45
Non empty interior, 58
Normal convex cone, 57
Normal integrand, 54

O
Optimal control, 17

P
Pettis controls, 35
Pettis integrable, 30
Pettis integration, 28–39
Pointwise converges, 29
(Ω)-Program, 103, 108
Property A, 136

Q
Quasi-concave, 136

R
Random norms, 76–79
Relatively compact, 32–33
Relaxed controls, 45
Re-simulation, 93–98

S
Scalarly derivable, 6
Second order differential game, 40
Separable Hilbert space, 40
Stochastic mesh, 76–79
Stochastic mesh methods, 61
Strategies, 39–58
Strictly concave function, 109
Subdifferential, 53
Sub-viscosity property, 41
Sweeping process (PSW), 39–58

T
Trajectory, 21
Trajectory solution, 20
Turnpike property (TP), 102

U
Uniformly integrable, 29
Uniqueness, 24
Upper–lower value function, 40
Upper semicontinuous function, 19, 102, 103

V
(υ)-good, 102, 105, 114, 121
(υ)-good admissible sequence, 102
(υ)-good program, 106, 110
(υ)-overtaking optimal program, 106, 110
Value function, 3
Variational problems with extended-valued integrands, 101
Viscosity, 3
Viscosity property, 25
Viscosity subsolution, 24–28, 53
von Neumann path, 102

W
Weak derivative, 6
Weakly optimal, 123
$W^{2,1}_{P,E}$ ([τ, 1])-solution, 37

Y
Young measures, 45

Advances in
MATHEMATICAL
ECONOMICS

Instructions for Authors

A. General

1. Papers submitted for publication will be considered only if they have not been and will not be published elsewhere without permission from the publisher and the Research Center for Mathematical Economics.

2. Every submitted paper will be subject to review. The names of reviewers will not be disclosed to the authors or to anybody not involved in the editorial process.

3. The authors are asked to transfer the copyright to their articles to Springer if and when these are accepted for publication.

The copyright covers the exclusive and unlimited rights to reproduce and distribute the article in any form of reproduction. It also covers translation rights for all languages and countries.

4. Manuscript must be written in English. Its pdf file should be submitted by e-mail: maruyama@econ.keio.ac.jp

Office of Advances in
Mathematical Economics
c/o Professor Toru Maruyama
Department of Economics
Keio University
2-15-45, Mita Minato-ku,
Tokyo 108-8345, JAPAN

B. Preparation of Manuscript

1. Manuscripts should be submitted in the pdf format. If this is not possible, two printouts of the manuscript must be submitted to the above postal address.

Manuscripts should be written in **LaTeX**. Please use Springer's LaTeX macro package (download from `ftp://ftp.springer.de/pub/tex/latex/svjour3/global/`). After acceptance, sending the original source (including all style files and figures) and a pdf (compiled output) are required.

Authors wishing to include figures, tables, or text passages that have already been published elsewhere are required to **obtain permission** from the copyright owner(s) for both the print and online format.

2. **The title page** should include:
- The name(s) of the author(s)
- A concise and informative title
- The affiliation(s) and address(es) of the author(s)
- The e-mail address, telephone and fax numbers of the corresponding author

Please provide an **abstract** less than 100 words. The abstract should not contain any undefined abbreviations or unspecified references.

Please provide 4 to 6 **keywords** which can be used for indexing purposes.

3. Please use the decimal system of **headings** with no more than three levels.

Abbreviations should be defined at first mention and used consistently thereafter.

Footnotes can be used to give additional information, which may include the citation of a reference included in the reference list. They should not consist solely of a reference citation, and they should never include the bibliographic details of a reference. They should also not contain any figures or tables. Footnotes to the text are numbered consecutively; those to tables should be indicated by superscript lower-case letters (or asterisks for significance values and other statistical data). Footnotes to the title or the authors of the article are not given reference symbols. Always use footnotes instead of endnotes.

4. **The Journal of Economic Literature index number (JEL classification)** should be indicated and the statement of the **2010 Mathematics Subject Classification (MSC) numbers** is desirable. You can check JEL classification with Internet at `http://ideas.repec.org/JEL/` as well as 2010 MSC numbers at `http://www.ams.org/msc`.

5. **Main text**: All **tables and figures** must be cited in the text and numbered consecutively with Arabic numerals according to the sequence in which they are cited.

For each table, please supply a table caption (title) explaining the components of the table. Identify any previously published material by giving the original source in the form of a reference at the end of the table caption. When preparing your tables and figures, size them to fit in the column width.

Short **equations** can be run in with the text. Equations that are displayed on a separate line should be numbered.

6. **Reference citations** in the text should be identified by numbers in square brackets. Some examples:
1. Negotiation research spans many disciplines [3].
2. This result was later contradicted by Becker and Seligman [5].
3. This effect has been widely studied [1–3, 7].

The list of references should only include works that are cited in the text and that have been published or accepted for publication. Personal communications and unpublished works should only be mentioned in the text. Do not use footnotes or endnotes as a substitute for a reference list. The entries in the list should be numbered consecutively.

•*Journal article*
Hamburger, C.: Quasimonotonicity, regularity and duality for nonlinear systems of partial differential equations. Ann. Mat. Pura. Appl. **169**, 321–354 (1995)
•*Article by DOI*
Sajti, C.L., Georgio, S., Khodorkovsky, V., Marine, W.: New nanohybrid materials for biophotonics, Appl. Phys. A (2007). doi:10.1007/s00339-007-4137-z
•*Book*
Geddes, K.O., Czapor, S.R., Labahn, G.: Algorithms for Computer Algebra. Kluwer, Boston (1992)
•*Book chapter*
Broy, M.: Software engineering - from auxiliary to key technologies. In: Broy, M., Denert, E. (eds.) Software Pioneers, pp. 10–13. Springer, Heidelberg (2002)
•*Online document*
Cartwright, J.: Big stars have weather too. IOP Publishing PhysicsWeb. `http://physicsweb.org/articles/news/11/6/16/1` (2007). Accessed 26 June 2007

Please use the standard abbreviation of a journal's name according to the ISSN List of Title Word Abbreviations, see `http://www.issn.org/2-22660-LTWA.php`

7. The purpose of the **author proof** is to check for typesetting or conversion errors and the completeness and accuracy of the text, tables and figures. Substantial changes in content, e.g., new results, corrected values, title and authorship, are not allowed without the approval of the Editor. After online publication, further changes can only be made in the form of an Erratum, which will be hyperlinked to the article.

8. Please use the standard **mathematical notation** for formulae, symbols etc.: Italic for single letters that denote mathematical constants, variables, and unknown quantities Roman/upright for numerals, operators, and punctuation, and commonly defined functions or abbreviations, e.g., cos, det, e or exp, lim, log, max, min, sin, tan, d (for derivative) Bold for vectors, tensors, and matrices.

Printed in the United States
By Bookmasters